FORENSIC BOTANY

PRINCIPLES AND APPLICATIONS TO CRIMINAL CASEWORK

FORENSIC BOTANY

PRINCIPLES AND APPLICATIONS TO CRIMINAL CASEWORK

EDITED BY
HEATHER MILLER COYLE

CRC PRESS

Boca Raton London New York Washington, D.C.

Library of Congress Cataloging-in-Publication Data

Forensic botany : principles and applications to criminal casework / edited by Heather Miller Coyle.
p. cm.
Includes bibliographical references (p.).
ISBN 0-8493-1529-8 (alk. paper)
1. Forensic botany. I. Coyle, Heather Miller.

QK46.5.F67F67 2004
363.25′62—dc22 2004051931
CIP

International Standard Book Number 0-8493-1529-8
Library of Congress Card Number 2004051931
Printed in the United States of America 1 2 3 4 5 6 7 8 9 0
Printed on acid-free paper

This book is dedicated to John, Lamorna, and Violet

"amor vincit omnia"

Foreword

Over the last 100 years, science has played an increasingly greater role in crime investigations. Unlike in popular TV shows, such as CSI, science may not always offer definitive solutions regarding a crime. It does, however, provide a special investigative role and in some scenarios can be quite revealing. Forensic science assists in solving crimes by characterizing physical evidence found at a crime scene for attribution purposes. In other words, scientific knowledge and technology are used as objective witnesses in solving the puzzle of who committed the crime. Forensic science is a combinatorial science that exploits diverse areas, which typically include the major disciplines of biology, chemistry, physics, and geology. Within each discipline are many subcategories of science that may be used in a forensic science investigation. For example, within biology some of the disciplines are medicine, pathology, molecular biology, immunology, odontology, serology, psychology, and entomology. The specific discipline(s) applied depends on the circumstances of the crime. Mathematics, especially statistics, is used, when appropriate, to place weight or significance on observations or data retrieved from crime scene evidence. The fields of human DNA analysis, hair morphology comparisons, handwriting, fingerprint identification, drug analysis, ballistics, tool marks, and others are well established in forensic science.

Over the past 20 years there has been a boon in the analysis of biological evidence heralded by the advent of DNA analysis. Myriad samples of human origin can be genetically characterized and often individualized. Any tissue containing nuclear or mitochondrial DNA, even minute and degraded samples, is now typeable. Moreover, this forensic biology subdivision has impacted substantially on the basic foundations in other forensic science fields, particularly in the areas of standard practices, quality assurance, validation, peer review, statistics, and education. Today, the biological sciences are the model of high quality in the forensic sciences.

Most, if not all, of the above-mentioned efforts in forensic biology have been dedicated to the identification of the source of human samples using molecular biology tools. However, other biological materials can be found associated with crimes. There are numerous examples where meaningful forensic evidence was derived from non-human biological samples for resolving crimes or discerning the cause of an accident or death. Hair and other tissues from domestic animals are often present as well as frequently transferred between individuals. The identification of poached endangered species by DNA has been used to develop evidence in the forensic wildlife arena. Verification of processed meat, caviar, seafood products, and so forth is a challenge where forensic biological sciences has played a role.

Microbial forensics, which entails the analysis of evidence from a bioterrorism act, biocrime, or inadvertent microorganism/toxin release for attribution purposes, is experiencing a huge influx of resources and efforts, particularly since the anthrax letters attack of 2001. The targets can be humans, agriculture, food and water, the environment, and our security and confidence. The challenges for identification are great and the number of species and the characterization of each species' diversity can be daunting.

One area of forensic biology that is often overlooked is botany. Botanical evidence has been identified in crimes for decades, if not centuries. Bulk and trace materials have been found on victims and suspects that associate them with crime scenes. Natural fibers such as cotton have provided exculpatory value. The pollen found contaminating evidentiary materials might be used for geolocation or seasonal associations. Foodstuffs found in the stomach have provided insight into the time of death. Some poisons derived from plants, such as hemlock, ricin, abrin, and nicotine, have been used as weapons in crimes.

Most traditional botanical evidence characterization is based on morphology, taxonomy, or isoenzyme typing, and cannot provide information to the level of individuality. There is nothing inherently wrong with such class characteristic evidence; it adds direct and/or circumstantial evidence for the fact finder to consider in the totality of the case. In general, class characteristics offer valuable information with regard to narrowing down the possible source or origin of specimens collected from the crime scene. Plant tissues differ from animal tissues in that the outer cellular membranes of plants are made up of components, such as lignin, that provide physical integrity to cellular structure and are very difficult to breakdown. Methods have been optimized to extract DNA for forensic analysis. Efforts have been dedicated to purify samples to remove phenolic substances, polysaccharides, and humic acids, which are often co-extracted with the DNA extracted from plants and can inhibit the PCR reaction.

Recently, the same benefits gained from exploiting the tools of the molecular biologist in human identification are beginning to be used for plant identification. PCR-based methods allow plant species identification while reducing the need for large quantities of starting material and strict preservation requirements. STR and VNTR typing and sequencing of mitochondrial DNA can be applied, as is done in human identification, to the genetic characterization of plants. DNA-based polymorphisms can also be detected in plants using randomly amplified polymorphic DNA (RAPD) analysis, arbitrary primed polymerase chain reaction (AP-PCR) typing, DNA amplification fingerprinting (DAF) systems, or amplified fragment length polymorphism (AFLP) analysis. In plants, another type of DNA exists in the cell, located in specialized organelles called chloroplasts. Chloroplasts are found in large numbers within some plant cells. The chloroplast membrane likely protects its DNA from microorganism degradation, making chloroplast genome analysis advantageous over tests based on nuclear DNA (as is experienced with mtDNA) for samples exposed to substantial environmental insult.

DNA typing analysis has been useful in a number of crimes such as identification of regulated or prohibited plant material, e.g., *Cannabis sativa* (marijuana), detection of fraudulent propagation of ornamental and agriculturally important cultivars, determination of illegal tree harvesting and association of a suspect with a victim buried under a tree in Arizona. Obviously, the identification and characterization of botanical materials can and have played a role in resolving some crimes and many analytical tools are currently available. It would be desirable to exploit such evidence when present (especially when other evidence is very limited in content or information or is not available). However, there are probably only a few of us in the forensic sciences with sufficient knowledge to begin exploiting botanical evidence. It is unlikely that the training and education effort experienced over the past few years on human identification will occur for plant identification. Therefore, a primer book was needed to familiarize the forensic scientist about the plant world. Dr. Coyle and her colleagues have produced such a book, and they should be commended for such an undertaking. They distill the material into easily readable text for the neophyte. Topics include: how humans rely on plants and the materials associated with plants; the basic biology of plants; plant databases; plant diversity; plant material as evidence and criminal investigations; molecular biology techniques for characterization of plants; and palynology. Also included are the general topics of basic genetics; development of forensic DNA analysis describing the human DNA typing experiencc; and legal considerations when applying new technologies.

Plant material is ubiquitous and thus is likely to be found in association with a number of crimes. The hope is that this book will serve as a springboard for the area of forensic botany, promulgate additional studies, and promote more efforts dedicated to exploiting other forms of biological evidence — evidence that can be used to exculpate those falsely associated with a crime as well as inculpate those who perpetrated the criminal act.

Dr. Bruce Budowle
FBI Laboratory
2501 Investigation Parkway
Quantico, VA 22135

Acknowledgments

Thank you to all the members of the Connecticut State Forensic Science Laboratory for their assistance and patience throughout the time I have been at the Forensic Science Laboratory. Without their humor and assistance, I would not have been able to get through the daily pressures of casework and the completion of this collaborative project.

A special thank you to the contributors of each chapter or for case examples: Dr. Henry C. Lee, Major Timothy Palmbach, Elaine Pagliaro, Dr. Gary Shutler, Dr. Robert Bever, Dr. Adrian Linacre, Dr. James Lee, Dr. Margaret Sanger, Dr. Kristina Shierenbeck, Dr. Carll Ladd, Eric Carita, Nicholas Yang, and Cheung-Lung Lee. Their outstanding expertise in their scientific fields and their dedication to writing a special topics chapter have resulted in a successful textbook that will be a reference guide to many scientists, attorneys, and law enforcement officials who may find themselves involved with a criminal case where botanical evidence may play an important role.

A special acknowledgment goes to Paul J. Penders for his professional photography and artistic presentation of plant samples. Rarely will a courtroom have the enjoyment of such pleasing images of botanical evidence, but it has been an exceptional delight to see plants portrayed in this textbook with such beauty and style. All of the professional plant photography is the work of Paul Penders unless otherwise noted in the photo caption.

Another special acknowledgement goes to the administration of the Connecticut State Forensic Science Laboratory for their encouragement and support of educational goals and the development of new scientific tools for use in the field of forensic science. In particular, I wish to thank Dr. Henry Lee for encouraging me to write often and write well and to Major Timothy Palmbach for providing encouragement and understanding.

And, finally, I wish to thank my parents — George and Marianne Miller — who have always encouraged me to reach my goals no matter what they may be, personally and professionally. Without their guidance, training, and support, I would not be the person I am today.

The Editor

Dr. Heather Miller Coyle is currently a lead criminalist for the Division of Scientific Services in Connecticut and specializes in DNA testing for the individualization of human and nonhuman evidence. In addition to her regular casework duties, Dr. Miller Coyle has focused on establishing a practical methodology for performing plant DNA typing on a variety of plant species, including *Cannabis sativa* (marijuana) for forensic cases in Connecticut, California, and Canada. Dr. Miller Coyle is currently engaged in several research collaborations with scientists in academia and private industry from the U.S., Croatia, and Taiwan.

Dr. Miller Coyle received her Ph.D. in plant biology from the University of New Hampshire in 1994 and performed postdoctoral DNA research at both Yale University and Boehringer Ingelheim Pharmaceuticals, Inc. She has lectured on forensics for numerous professional organizations and has taught forensic DNA courses at Wesleyan University and the University of Connecticut, and is an adjunct faculty member at the University of New Haven. Dr. Miller Coyle has authored many scientific articles in professional journals and has contributed to several books. She has testified as an expert witness in forensic biology and DNA for the State of Connecticut and consulted on cases for Bermuda and Illinois.

Residing in New York State, Heather and her husband John, an officer in the Fire Department of New York City, have been married for 8 years and have two young daughters.

Contributors

Robert A. Bever
The Bode Technology Group, Inc.
Springfield, Virginia, USA

Vaughn M. Bryant, Jr.
Palynology Laboratory
Department of Anthropology
Texas A&M University
College Station, Texas, USA

Eric Carita
Center For Non-Traditional DNA
Typing Methods
University of Connecticut
Storrs, Connecticut, USA

Heather Miller Coyle
Department of Public Safety
Division of Scientific Services
Meriden, Connecticut, USA

Hsing-Mei Hsieh
Department of Forensic Science
Central Police University
Kwei-San, Taoyuan, Taiwan

Ian C. Hsu
Department of Nuclear Science
National Tsing Hua University
Hsin-Chu, Taiwan

Carll Ladd
Department of Public Safety
Division of Forensic Science Services
Meriden, Connecticut, USA

Cheng-Lung Lee
National Tsing Hua University
Hsin-Chu Municipal Police Bureau
Hsin-Chu, Taiwan

Henry C. Lee
University of New Haven
Henry C. Lee Institute of Forensic Sciences
West Haven, Connecticut, USA

James Chun-I Lee
Department of Forensic Science
Central Police University
Kwei-San, Taoyuan, Taiwan

Adrian Linacre
Forensic Science Unit
Department of Pure and Applied Chemistry
University of Strathclyde
Glasgow, United Kingdom

Dallas C. Mildenhall
Institute of Geological and Nuclear Sciences
Lower Hutt, New Zealand

Lynne A. Milne
School of Earth and Geographical Sciences
and Centre for Forensic Science
The University of Western Australia
Crawley, WA, Australia

Elaine Pagliaro
Department of Public Safety
Division of Scientific Services
Meriden, Connecticut, USA

Timothy Palmbach
Department of Public Safety
Division of Scientific Services
Meriden, Connecticut, USA

Paul J. Penders
Department of Public Safety
Division of Scientific Services
Meriden, Connecticut, USA

Margaret Sanger
Appalachian H.I.D.T.A. Marijuana Signature Laboratory
Frankfort, Kentucky, USA

Kristina A. Schierenbeck
Department of Biological Sciences
California State University
Chico, California, USA

Gary Shutler
Washington State Patrol
Crime Laboratory Division
Seattle, Washington, USA

Nicholas CS Yang
Department of Public Safety
Division of Scientific Services
Meriden, Connecticut, USA

Contents

Introduction to Forensic Botany

HEATHER MILLER COYLE

Contents

1.1 Plants Are Ubiquitous

Plants are ubiquitous in nature, essential for all human and animal existence. They are critical to the earth's atmosphere and to other forms of life, and serve a function as intermediaries by converting solar energy into complex molecules. While some plants provide a source of food, others provide fiber, medicine, and aesthetic pleasure. The average person enjoys plants in a multitude of ways throughout the day. The pure green of the golf course, the sweet smell of a rose, the calming effect of chamomile tea, a glass of wine from a superior grape cultivar, and the comfort of well-worn cotton blue jeans are just a few examples.

0-8493-1529-8/05/$0.00+$1.50

These are common, pleasant examples of plant usage in our society, but what happens when plants become associated with criminal activities?

1.2 Plants as Poisons

Many homicides, accidental deaths, and suicides have occurred through the inappropriate and deliberate use of plants to achieve harmful effects. Many substances, including common table salt, can be considered poisonous if consumed in sufficient quantity. The use of poison to commit murder is more prevalent with women, but plenty of documented cases exist for both sexes. The list of plants that have been used for both deliberate and accidental death is lengthy. A few examples include the following: hemlock, wolfsbane (monkshood), belladonna (deadly nightshade), tobacco (nicotine), nux vomica tree (strychnine), foxglove (digitalis), and castor bean (ricin). The early nineteenth century was the "fashionable period" for murder by poisoning. In fact, poisoning was so common that many people employed food tasters to sample their food before ingestion as a safeguard to their health.

1.3 Plants as Trace/Transfer Evidence

In addition to the toxic properties of some plants, botanical samples can serve as key pieces of trace evidence. Forensic botany is a subdiscipline of forensics that is still, perhaps, in its infancy. It is, however, becoming a more common tool as law enforcement agencies, forensic scientists, and attorneys become more familiar with the applications of plant material to casework. Also, as plant research continues in academic and private research laboratories, new tools are being developed that can be eventually applied to forensic botanical evidence to aid in criminal and civil cases. In addition, these new tools can be important for national security and historical issues. Plant evidence is commonly associated with crime scenes and victims; typically, stems, leaves, seeds, flowers, and other identifiable parts of plants are useful as trace botanicals. One might imagine that the chemical composition of herbs, spices, condiments, beverages, fibers, medicines, plant resins, and oils, for example, may be useful for forensically relevant plant "signature" programs just as those used in drug chemistry (e.g., cocaine, heroin) if appropriate databases can be constructed.

1.4 Forensic Botany

Forensic botany is a marriage of many disciplines and results ultimately in their application to matters of law. The botanical aspects of forensic botany include plant anatomy, plant growth and behavior, plant reproductive cycles and population dynamics, and plant classification schemes (morphological and genetic) for species identification. The forensic aspects require an understanding of what is necessary for botanical evidence to be accepted as evidence in our judicial system. Forensics requires recognition of pertinent evidence at a crime scene, appropriate collection and preservation of evidentiary material, maintenance of a chain of custody, an understanding of scientific testing methods, validation of new forensic techniques, and admissibility criteria for court. The goal of this textbook is to provide basic knowledge to anyone wishing to study or utilize botanical evidence in forensic casework. The framework for plant biology and general forensics is described in the first few chapters of this textbook. After providing a scientific foundation for the field of forensic botany, we describe applications and casework examples in detail. A wide variety of forensic botany examples include the use of several plant parts (leaves, flowers, pollen, wood) and plant species. Almost as varied are the scientific methods utilized in forensic botany, and these methods range from simple techniques (e.g., light microscopy) to more technical molecular biology techniques (e.g., DNA sequencing). Plants have been used as evidence in criminal cases for kidnapping, child abuse, hit-and-run motor vehicle accidents, drug enforcement, homicide, sexual and physical assault, the establishment of time of death, and verification of an alibi. In addition, new applications are under development to use plant material in forensics as "tracers" to aid in the identification of missing persons, to track drug distribution patterns, and to link bodies to primary crime scene locations after they have been dumped at secondary sites.

1.5 Plants in Our Society

To understand the widespread application and potential utility of plants in forensics, we discuss a few brief examples of plant usage in human society. As these examples are presented, consider the number of plant-based items that may be found on your person, among your private possessions, and in your home and workplace — and consider, one day, that they may be useful as critical trace evidence.

1.5.1 Food

Apples are generally considered to be a wholesome, healthful addition to the daily diet as a good source of vitamins and fiber (Figure 1.1). In actuality, this concept was promoted by the apple industry in response to the renouncement of apples by Carry Nation as part of the Prohibition Act. The Women's Christian Temperance Union was opposed to apples because they were, in

Figure 1.1 Apples are an edible mainstay of many people's diets and have been successfully swabbed for suspect DNA profiles after burglaries.

part, responsible for alcohol use on the frontier. In the early 1900s, the apple industry began promoting the healthful benefits of apples, and today, we have many apple cultivars to choose from at our markets. Interestingly, burglars often sample fruit and other foods in the homes they are invading while pilfering goods.

1.5.2 Fiber

One prevalent clothing fiber in our society is cotton. Cotton comes from the elongated epidermal hairs on the seeds of the *Gossypium hirsutum* plant. Prior to mechanized harvesting of cotton fibers, flax (*Linum usitatissimum*) was the most common fiber plant for the textile industry. Flax fibers, unlike cotton, are sclerenchyma fibers from the stem of the flax plant. Flax fibers are commonly woven into linen, and certain cultivars are used to produce cigarette papers and linseed oil. As trace evidence, the source of clothing, carpet fibers, rope, twine, and threads can be useful for associating a victim to a suspect or individuals back to a primary crime scene for an investigative lead.

1.5.3 Medicine

Herbal remedies and folklore investigations to identify active chemical components have long been part of human culture and have played an important role in the discovery of useful medicinal compounds (e.g., aspirin). In addition, well-preserved stomach contents from ancient human remains have yielded insight into rituals involving herbs and food. For example, in 1984, a peat cutter near Manchester, England, discovered a well-preserved human leg and called in the police to investigate. The body of Lindow Man was recovered, and radiocarbon dated to approximately A.D. 50–100, and he was determined to be a member of the Celtic tribes. The archeobotanists used electron spin resonance (ESR) applications to examine the gut contents of the preserved man and discovered burned, unleavened barley bread and put forth the concept of a sacrificial victim during the Feast of Beltain. This feast was celebrated on May 1 as a pagan ritual, and it is thought that the unlucky person to choose the blackened bread was designated as the sacrifice.

1.5.4 Beauty

Plants are key components in many herbal shampoos, soaps, cosmetics, and perfumes. Not only are botanicals used in human cosmetics and have appealing scents, but they beautify our environment as well. The *Dahlia* flower, for example, has an unusual history (Figure 1.2). Originating from Mexico and called *Cocoxochitl* by the Aztec Indians, it was discovered and seeds were sent to a French priest studying botany in Madrid, Spain. It was then renamed

Figure 1.2 The dahlia is one example of a common horticultural plant found in landscaping across the U.S. Pollen from flowers can often be useful for providing additional evidence if found on suspect clothing.

after Dr. Dahl, a student of botany, who was interested in its tuberous roots as a potential food source rather than for the showy blossoms that gardeners covet today. After planting in English greenhouses, most of the tubers rotted away until the dahlia was improved by breeding efforts in the late 1800s and became accessible to gardeners everywhere. The American Dahlia Society recognizes 12 groups of cultivars based on the head morphology and the flowers, and suggests that most modern cultivars were derived from the parental *Dahlia pinnata* and *Dahlia coccinea.*

1.5.5 Recreation

The American obsession with green lawns can be visualized in the numerous golf courses, city parks, and extensive front and backyards that are ingrained parts of suburban life. Prior to the Civil War, few Americans had lawns. However, in the 1950s, the term "lawn" was used in reference to a portion of land kept closely mown in front or around a house. In the 1950s, turfgrass breeding programs gave rise to several new grass varieties that offered improved heat, drought, and disease resistance. In the 1970s and '80s, lawn specialists began recommending blends of grasses (e.g., fescue, bluegrass, perennial ryegrass) rather than a monoculture of a single grass species. Today, American homeowners spend enormous capital and energy on achieving a perfectly groomed, green lawn as a setting for their homes. In fact, an entire lawn-care industry has developed around this particular aspect of suburban homes. Grass samples may be one of the most abundant types of botanical evidence found at crime scenes simply due to the American obsession with the lawn.

1.5.6 Law Enforcement

Plants may be present as biological evidence in many ways:

- Seeds caught and carried in a pant cuff
- Grass stains on a dress after a sexual assault
- Plant leaves and stems snagged and carried in a vehicle's undercarriage, grill, wheel wells, hood, or trunk
- Stomach contents with vegetable matter to aid in verification of an alibi
- Use of pollen to date the burial of skeletal remains in a mass grave

All of these examples and more can assist the forensic community in associating a person to an object, a person to a crime scene, or a suspect to a victim. "Every criminal leaves a 'trace' (evidence)" is a phrase with some accuracy. That "trace" may very well be biological plant material. The remaining chapters will illustrate important concepts in plant biology, key aspects of forensic science, new technological uses for plant DNA, and the legal considerations for using botanical evidence in court.

General References

Bailey, L.H. and Bailey, E.Z., *Hortus Third: A Concise Dictionary of Plants Cultivated in the United States and Canada*, Macmillan Publishing Co., New York, 1976, 360–61.

Pollan, M., *The Botany of Desire: A Plant's-Eye View of the World*, Random House, New York, 2001, 1–58.

Ross, A. and Robins, D., *The Life and Death of a Druid Prince*, Summit Books, New York, 1989, 9–14.

Scott Jenkins, V., *The Lawn: A History of an American Obsession*, Smithsonian Institution Press, Washington, D.C., 1994, 2.

Whittle, T., *The Plant Hunters*, PAJ Publications, New York, 1988, 5–6.

Basic Plant Biology

HEATHER MILLER COYLE

Contents

0-8493-1529-8/05/$0.00+$1.50

2.1 What Is a Plant?

A plant may be defined as any member of the kingdom Plantae typically characterized by the following:

- A lack of locomotion
- The absence of sensory organs or a nervous system
- The possession of a cellulose cell wall
- A nutrition system in which carbohydrates are formed through photosynthesis
- The alternation of sexual versus asexual generations

Plants can be large or small, flat or erect, flowering or nonflowering, green or possessing other pigmentation. The kingdom Plantae is enormously varied in physical appearance and in growth habits. Most plant groups include members containing chlorophyll, a group of pigments having the vital function of absorbing light energy for photosynthesis. *Photosynthesis* is the process by which plants convert carbon dioxide and water into organic chemicals using the energy of sunlight. Organisms commonly classified as plants include the following:

- Prokaryotic organisms
  Blue-green algae
  Bacteria (some species)

- Eukaryotic organisms
  Algae: red, green, brown, golden-brown
  Euglenoids
  Diatoms
  Dinoflagellates
  Molds: water, slime, bread
  Chytrids
  Fungi: club, sac, imperfect
  Lichens
  Mosses
  Hornworts
  Liverworts
  Ferns

Horsetails
Gymnosperms: ginkgos, cycads, conifers
Angiosperms: all flowering plants

Organisms have different modes of nutrition and can be classified as autotrophs, heterotrophs, photoautotrophs, or chemoautotrophs. An autotroph is sometimes called a "self-feeder," which may give the impression that it is a self-sustaining organism. In actuality, this means that autotrophs obtain nutrients differently; they do not eat or decompose other organisms in order to survive. Plants are photoautotrophs and use sunlight for energy to manufacture lipids, proteins, carbohydrates, and other necessary organic substances. Chemoautotrophs are a rare form of self-feeding bacteria that obtain their energy by oxidizing inorganic substances (e.g., sulfur). Heterotrophic organisms are not self-sufficient in the sense that they rely on other plants, insects, or animals for their nutrition (e.g., humans). Heterotrophic organisms can be further subdivided into herbivores (plant-eaters), carnivores (meat-eaters), insectivores (insect-eaters), and omnivores (indiscriminate eaters of plants, animals, and insects). Heterotrophs include some fungi and bacteria that decompose substrates in order to find the necessary nutrition.

2.2 The Process of Photosynthesis

The process of photosynthesis sets plants apart from animals. While animals must seek their own nutrition by either grazing on plant life or participating in a predatory act, plants are able to remain in one general location and will convert light energy into usable nutrition. Photosynthesis is a complex process but can be described in the simpler terms of the following equation:

$$6\ CO_2 + 6\ H_2O + \text{light energy}$$

is converted in the presence of chlorophyll and enzymes to yield $C_6H_{12}O_6 + 6\ O_2$. Other molecules are produced as by-products of this process, including organic phosphates, organic acids, and some amino acids.

Light energy, also called radiation, is a form of electromagnetic energy that travels in waves. One wavelength is the measurable distance between the crest of two waves. The entire electromagnetic range of energy is called the electromagnetic spectrum; however, only a small portion is critical to plants. This small portion is called the visible light spectrum, which is in the range of 380 to 750 nm (see Figure 2.1). We perceive different visible light wavelengths as different colors (e.g., red, blue, yellow). Photosynthetic plants utilize wavelengths of light in the red and violet-blue areas of the visible light spectrum because the photosynthetic pigments capture those particular

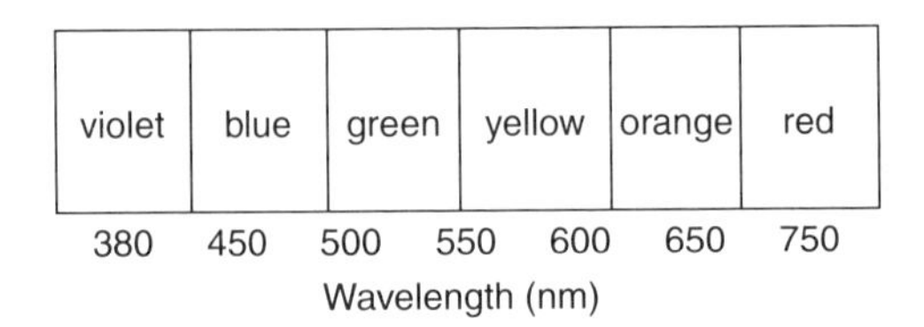

Figure 2.1 Visible light travels through the air as energy waves of different lengths. The different wavelengths of light are perceived by our eyes as colors. Red has the longest wavelength, violet has a shorter wavelength, and the other colors fall between red and violet. Visible light is required for the process of photosynthesis.

wavelengths optimally. Many plants appear green to the human eye because plants reflect rather than absorb the wavelength for "green" light. Chlorophyll molecules act in combination as "light traps" in the chloroplast membrane to capture photons (energy light units). During the daytime, abundant light is available, and plants capture photons and convert them to intermediate compounds (ATP, $NADPH_2$) in the light-dependent phase of photosynthesis. Carbon dioxide fixation is not a light-dependent process and can occur by multiple biochemical pathways. C3 pathway plants produce a three-carbon compound, phosphoglyceric acid, as the first product in carbon dioxide fixation. C4 pathway plants produce a four-carbon oxaloacetic acid as the first product in the pathway. A third form of carbon dioxide fixation performed by some succulent plants is called the CAM pathway. The evolution of C4 and CAM pathways is an adaptive strategy to maintaining photosynthesis during hot, arid conditions with stomata only partially open or closed to reduce water loss. The final purpose of carbon dioxide fixation is to use CO_2, ATP, and $NADPH_2$ to produce simple sugars for nutrition (see Figure 2.2).

2.3 General Plant Classification Schemes

Plants are classified into different phylogenetic groups; higher and lower plants, algae and land plants, and vascular and nonvascular plants are a few examples of classification schemes. The focus of this book will primarily be on vascular plants as the majority of species belong to this group. Higher plants can conduct water and nutrients (vascular plants or tracheophytes) through a specialized system of cells. Within the plant kingdom, vascular plants include four phyla:

- Sphenopsida (horsetails)
- Lycopsida (club mosses)
- Psilopsida (primarily fossils)
- Pteropsida (ferns, gymnosperms, and angiosperms)

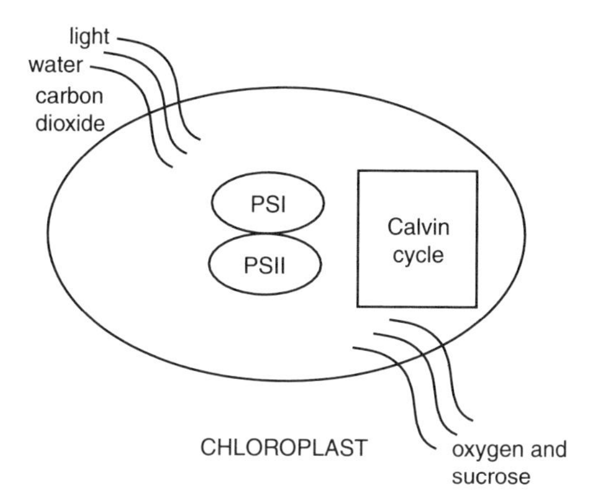

Figure 2.2 The biochemical reactions of photosynthesis are complex and are represented here in a simplified diagram. Light energy is received by the chlorophyll molecule in the plant chloroplast, and this converts the chlorophyll to an excited state. The excited electron energy is passed to a neighboring molecule and enters into a biochemical cycle that ultimately converts light into sugars and other organic compounds that are used to maintain the plant.

Those plants that reproduce by seed are further categorized as flowering plants (angiosperms) or cone-bearing plants (gymnosperms). There are a wide variety of classification schemes for plants. Some are based on physical appearance, environmental growth habitat, reproductive strategy, morphology of specific plant structures, and evolutionary relationships between segments of DNA, to illustrate a few possible schemes. A separate chapter of this book will be devoted to a complete description of phylogenetic analyses for the kingdom Plantae.

2.4 Plant Survival Strategies

Higher plants and animals are morphologically and structurally different but at the cellular levels are remarkably similar, arguing for a common ancestral origin. Plants have a specific set of problems, however. Nonlocomotion or immobility means plants are confined to a constrained geographic space and must access all necessary nutrients and resources from a limited space. This has necessitated that they develop some unique strategies to respond to their problems. The ability to convert radiant sunlight into energy has been an adaptation to accommodate the lack of mobility of most land plants. The development of modified carbon fixation pathways has increased the efficiency in which some plants are able to use limited water resources. In addition, the inability to escape predators has resulted in a striking number of defense mechanisms for plants, including thorny stems and leaves to reduce palatability for herbivores. The chemical production of phenolics, tannins, and resins that also have undesirable tastes reduce herbivore damage to plants. When damaged by other animals or insects, many plant species can regenerate their tissues to heal the injury. Also, reproductive strategies for plants are numerous and varied to aid in the survival of the species. Such strategies may include a rapid generation time, production of high numbers of progeny, wind and insect pollen dispersal mechanisms, and modes of wind and animal seed dispersal.

2.5 Subspecialties of Forensic Botany

When studying a new field of science, one must become familiar with the common words and phrases used to describe various aspects of forensic botany. A glossary of terms is listed in Appendix B. The following is a list of subspecialties and descriptions with which a student of forensic botany should become familiar.

2.5.1 Plant Morphology

Plant morphology can be defined as the study of the external structure of plants. The external structures can be useful for species identification. Certain shapes of leaves, hair characteristics, spacing of stomata, and specialized glands called trichomes can aid in the classification of a plant species. In addition, these features can be useful when making a physical match between two leaf fragments to show a continuity of patterns on leaf surfaces. Typically, a microscope with good resolution (100 to 400X magnifications) can be used to study these plant features.

2.5.2 Plant Anatomy

The study of the internal structure of plants is called *plant anatomy.* When a stem, root, or leaf is dissected, the cells can be arranged in specific patterns that may be useful for classification and identification. For example, the internal arrangement of cells in the root structure of a dicotyledonous (the first true leaves of a seedling occur in pairs) versus a monocotyledous (the first true leaf is a single leaf) plant is characteristic.

2.5.3 Plant Systematics

The study of plant diversity that includes taxonomy and plant evolutionary relationships is referred to as *plant systematics.* This is the study of how plants are currently classified or grouped based on different morphological or molecular characteristics. A basic understanding of how different plant species are related to one another can help with distinguishing between plant species that look similar (and may be closely related) and accurately identifying real differences between two samples that may have evidentiary value from a crime scene.

2.5.4 Palynology

The study of plant pollen and species identification based on pollen characteristics is *palynology.* Pollen from different flowering plant species can be useful in forensic casework as trace evidence, especially since it can provide investigative leads for the timing of deposit and for making geographic inferences. Photoperiod and climate can dictate the development and release of pollen; thus, an association of pollen with skeletal remains, for example, can provide information related to the annual season in which death may have occurred.

2.5.5 Plant ecology

The use of plant growth patterns and habitat to infer information regarding geographical restrictions for plant population distribution patterns is part of *plant ecology.* This area of study can also be useful in forensics to gauge how long skeletal remains have been in a specific location based on the extent of moss or fungal growth on the bones, or by the age of a young sapling tree growing through the eye socket of a skull.

2.5.6 Limnology

The study of aquatic plant life from freshwater lakes, ponds, and streams is called *limnology.* Limnology may be useful for associating victim and suspect to a general location based on the algal composition on clothing or shoes, for example, or for indicating freshwater drowning based on the presence or absence of diatoms.

2.6 Plant Architecture

2.6.1 Roots

Plants are comprised of roots, stems, leaves, and flowers. Roots anchor the plant in the soil and take up minerals, salts, and water, and store food. In grasses and monocotyledonous species, the roots form a fibrous cluster and are of approximately equal size. In most dicotyledonous plant species, like carrots and radishes, a main root (taproot) grows downward and smaller roots grow off the main root. The smaller roots and root hairs that branch off the main root function to increase the surface area of the root for better absorption of nutrients and water from the soil (see Figure 2.3).

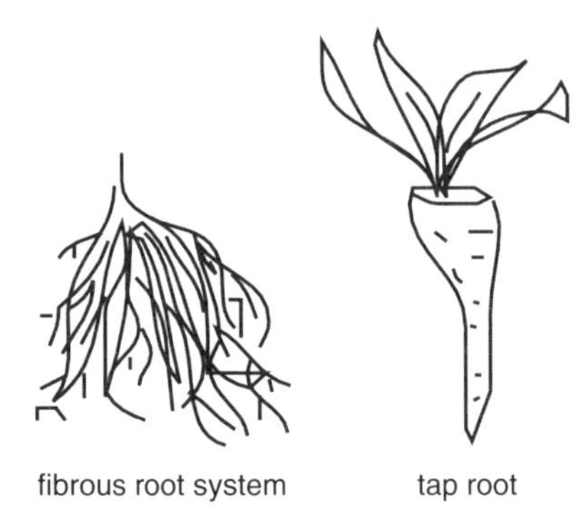

Figure 2.3 Different types of root systems are found on plants. A taproot consists of one main root with smaller lateral roots growing on the sides of the main root. A fibrous root system has smaller, entangled roots with no central root.

2.6.2 Stems

Stems are support structures above the soil surface and can be green or woody, and can possess protective thorns or modifications such as tendrils. Leaves are distributed in regular patterns along a stem. The part of the stem where a leaf is connected is a node; the stem between nodes is called the internode. If a stem is cut as a cross section, the shape and pattern of the vascular bundles can be useful features to aid in the classification and identification of a plant species.

2.6.3 Flowers

Floral buds can be arranged in clusters along a stem, between the node and the stem (an axillary position), or at the tip of each stem (a terminal position) (see Figure 2.4). A flower consists of specialized leaves, and those parts are

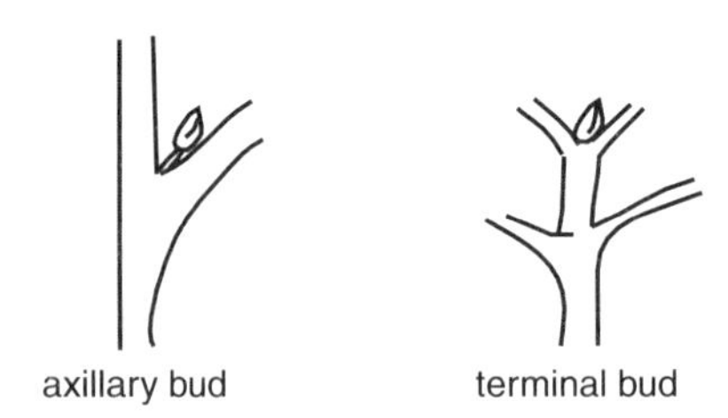

Figure 2.4 The arrangement of flowers on a stem can be a useful feature for distinguishing the plant species in question. Flowers that grow from the axil of the leaf are termed axillary. When arranged in a circle around the stem, they are called whorled. A raceme is a cluster of flowers with each arranged on its own common stem. Other forms of flower arrangements include umbel (clustered like the spreading ribs of an umbrella), head (a clustered circular flower like a dandelion), or a spike (flowers cluster close to the stem along a lengthy axis).

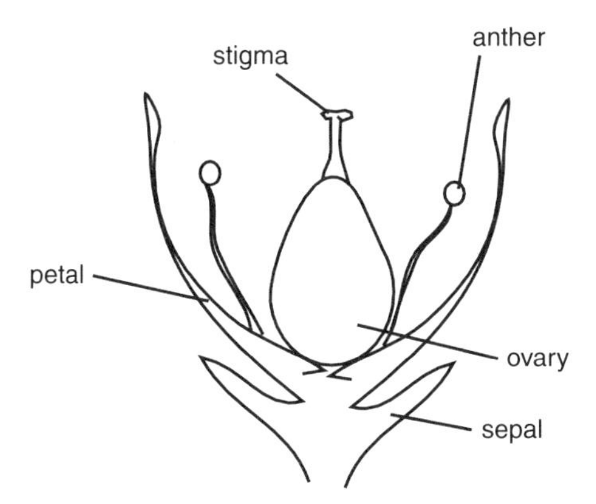

Figure 2.5 A schematic cross-sectional view of a typical perfect (having both male and female reproductive organs) flower. The female organs are the stigma, style, and ovary (where the ovules are located for fertilization); the male organs are the anthers (where the pollen is located) and filaments.

called the sepals, petals, stamens, and carpels. The sepals combined are called the calyx; the petals combined are called the corolla. The stamens are comprised of filament and an anther; the anther is the organ where pollen grains develop and mature. The carpel(s) make up a structure called the pistil. The bottom of the pistil is the ovary where female eggs ("seeds") develop from ovules. The middle portion of the pistil is the style; the top part is called the stigma (see Figure 2.5).

2.6.4 Leaves

As most people think of them, leaves are specialized structures for performing photosynthesis and transpiration. Photosynthesis, as previously discussed, is a process for obtaining nutrients for a plant. Transpiration is a process by which plants lose water to the environment as vapor. The high surface-to-volume ratio of a flat, thin leaf is helpful for gas exchange as well as allowing for sunlight to penetrate all cells. In the epidermis (the surface layer of cells), stomata function for gas exchange between leaf tissues and the atmosphere. Single stomata are called a stoma, and it consists of an opening with two guard cells surrounding it to assist in regulation of gas exchange.

2.7 Practical Plant Classification Schemes

A practical plant classification scheme differs from those used by traditional botanists to group plants based on leaf morphology or designated Latin nomenclature. For the crime scene investigator, some simple groupings based on practical usage of plants can be helpful. Such groupings could be vegetable crops, spices, common ornamentals and landscape plants, floral bouquets, common food items, and plant-based poisons and drugs.

2.7.1 Vegetables and Herbs

Included in this group would be intact plants, seeds, and fruits of the following:

- Tomato
- Potato
- Beans
- Peppers
- Eggplant
- Corn
- Squash
- Pumpkin
- Cucumber
- Lettuce
- Broccoli
- Cauliflower
- Peas
- Garlic
- Onions
- Rhubarb

Many avid gardeners also grow and harvest their own herbs, such as

- Pennyroyal
- Sage
- Parsley
- Rosemary
- Bee balm
- Oregano
- Lavender
- Thyme
- Coriander
- Mint

These herbal items may be found fresh or dried, and the types of spices that are grown in a particular region of the U.S. are highly dependent on the climate. At crime scenes, one would expect food items to be fairly common, and these items may be useful for establishing or refuting alibis or providing linkages of bodies (through stomach content analysis) to primary crime scenes.

2.7.2 Fruit-Bearing Trees and Plants

Apples and crab apples, pears, plums, peaches, oranges, grapefruit, lemons, kumquats, strawberries, raspberries, and blueberries are all common examples that may be regular household items or garden-grown plants depending on the geographic region of the U.S. Additional specialty fruits are grown primarily as interesting ornamental trees and shrubs. These specialty plants are typically ordered through a limited number of suppliers throughout the U.S. If some of these plant and tree species are associated with a crime scene, their inherent atypical geographic distribution and rarity can make them more valuable or probative as forensic botanical evidence. Some examples of unusual fruit-bearing trees and shrubs are

- Pawpaw
- Gooseberry
- Nanking cherry
- Medlar
- Juneberry
- Maypop
- Kaki and American persimmon
- Raisin tree
- Currants (red, white, or black)
- Elaeagnus
- Actinidia
- Jujube
- Alpine and musk strawberries
- Mulberry
- Lowbush blueberry
- Asian Pear
- Jostaberry
- Kiwis
- Cornelian cherry

2.7.3 Landscaping Plants: Trees, Shrubs, and Vines

Some relatively common plants found around the exterior of a home, condo, townhouse, restaurant, and so forth include the following:

- Azaleas
- Rhododendrons
- Maples
- Japanese maples
- Yews
- Holly

- Arborvitae
- Balsam fir
- Pine species
- Cedar
- Buckeye
- Oak species
- Dogwood
- Mountain ash
- Lilacs
- Spruce species
- Clematis
- Ivy species
- Hybrid tea roses
- Chrysanthemums
- Daylilies
- Hostas
- Tulips
- Daffodils
- Crocus
- Trillium

2.7.4 Grasses

Grass species are tremendously abundant, and many species are difficult to distinguish from one another unless one is a botanical expert. In the Northeastern region alone, over two hundred different native grass species have been characterized. Broad classifications for practical purposes could include

Ornamental landscape grasses: Bamboo, pampas grass, uva grass, Chinese silvergrass, ribbon grass, and Job's tears
Lawn grass species: Fescues, Bermuda grass, redtop, bentgrass, rye grass, centipede grass, and *Zoysia* species
Cultivated crop grass species: Barleys, rice, wheat, oats
Native (noncultivated) grass species: Buffalo grass, blue grama grass, june grass, needle grass, sedges, bulrushes, and rushes

All grasses belong to the family Gramineae and through agriculture have become the mainstays of our society's required food and nutrition.

2.7.5 Common Poisonous Plants

Plants can be poisonous through direct physical contact or from chemical irritation (e.g., stinging nettles) or an allergic reactive dermatitis (e.g., poison oak, poison ivy). Direct physical injury from a plant is typically through the action of spines, thorns, prickles, barbs, awns, or serrated leaf edges. Physical injury from these plant parts is temporary and nonlife threatening. Many plants produce chemicals that act as irritants of human eyes and skin. These include the following:

- Buttercups (*Ranunculus*)
- Mums (*Chrysanthemum*)
- Maidenhair tree fruit (*Ginkgo biloba*)
- Spurges (*Euphorbia*)
- Manchineel (*Hippomane mancinella*)

Plants that have stinging hairs produce both a chemical irritation and mechanical injury. Some examples are stinging nettle (*Urtica*), horse nettle (*Solanum*), and some primroses (*Primula*). Plants that result in a systemic poisoning and sometimes death are true poisons. The symptoms of poisoning are typically unpleasant and painful, and may include convulsions, hallucinations, nervous system disorders, loss of coordination and respiratory abilities, and coma. Death by poison can arise from four general possibilities: home medicine therapy (herbal supplements), search for native edible plant species (mistaken identification), accidental ingestion, and intentional poisoning (assault or homicide). Some home-brewed medicinal therapies include mild doses of belladonna, tincture of opium, spirits of camphor, and quinine. However, one of the great difficulties with herbal supplements and home remedies made from plant products is that each plant, in different quantities or potency, produces the active compounds. This makes it difficult in practice to administer an accurate dose to the patient. Some factors that may affect the production of medicinal plant compounds include differential gene regulation, environmental conditions (e.g., soil composition, photoperiod), and timing of harvest (plant maturity or growth phase). A survey of South African cases from a Johannesburg forensic database indicated that 206 cases between 1991 and 1995 had a traditional medicinal remedy stated as either the cause of death or associated with an unexplained death. Herbal plant material was associated with 43% of those cases. In France, 40 plants have been characterized for active compounds associated with fatal cases. Procedures for the detection of these compounds from whole blood by high-performance liquid chromatography–(tandem) mass spectrometry have been developed.

The typical example of death by mistaken identification of an edible plant species is the misidentification of mushrooms and fungi (Table 2.1). In addition, many plant species contain alkaloids, and it is very common for children, in particular, to accidentally overdose and die from ingestion of berries or leaves. Some examples of such plant species include *Solanum*

Table 2.1 Common Mushroom Toxins

Toxin	Genus	Description
Amatoxins	*Amanita, Conocybe, Galerina, Lepiota*	50% fatality rate, fatal dose approximately 2 ounces
Gyromitrin	*Gyromitra, Ascomycetes*	Hydrolytic compound monomethylhydrazine used in manufacturing rocket fuel; narrow threshold between asymptomatic and severe poisoning
Ibotenic acid (muscimol)	*Amanita*	Fatal in large doses
Psilocybin (psilocin)	*Psilocybe, Panaeolus, Conocybe, Gymnopilus*	Psychedelic properties, anxiety
Toxin unknown	*Paxillus*	Toxic when raw, edible if cooked; hypersensitivity response may result in sudden death
Toxin unknown	*Cortinarius*	Liver and kidney damage, up to 3 weeks post-ingestion

dulcamara (common nightshade or European bittersweet) and *Aesculus hippocastanum* (horse chestnut). *S. dulcamara* has bright, red-orange berries that are appealing to children for decorations, pretend "tea parties," etc., and toxicity results from the action of the chemical compound solanine, an alkaloid. With *A. hippocastanum* poisoning, the toxic ingredient is saponin, which is especially damaging when repeatedly ingested. Other plants that contain saponin are commonly known as

- Blue cohosh
- Pigeonberry
- Skyflower
- English ivy
- Yam bean
- Poke weed
- Rain tree
- Soapberry
- Cow cockle
- Mock orange

2.7.6 Types of Plant Toxins

Many plants can cause an allergic response, which is a result of an immune response. When an antigen (allergy-producing factor) is introduced to the body, antibodies (antigen-attacking factors) may be produced. This antigen–antibody interaction then stimulates the body to produce certain substances (e.g., histamine) that can harm cells in a localized or systemic manner. This is a natural response designed to protect the body from invading pathogens (e.g., disease-causing bacteria, fungi, viruses). Direct contact with a plant (e.g., poison ivy) or inhalation of pollen (e.g., hay fever) can result in redness of the eyes, irritation of the mucous membranes, nausea, difficulty breathing, and swelling or redness of localized regions of the skin. Photodermatitis is a sensitivity to sunlight after exposure to certain plant substances to wet skin. Some plants that can cause the redness and blistering common to photodermatitis include celery (*Apium graveolens*), wild chervil (*Chaerophyllum bulbosum*), and carrot (*Daucus carota*).

Poison ivy and posion oak (*Rhus* or *Toxicodendron* species) are common plants that can cause contact dermatitis due to a delayed hypersensitivity response to the chemical compound urushiol (3-n-pentadecylcatechol). An individual's sensitivity to urushiol can vary greatly and may result in a mildly irritating skin rash or become extreme to the point of hospitalization. Nine plant families contain the primary culprits of allergic contact dermatitis:

- Anacardiaceae (poison ivy, poison oak, mango, cashew)
- Amaryllidaceae (narcissus, daffodil)
- Asteraceae (dog fennel, daisies, chrysanthemums)
- Poaceae (cereals, grasses)
- Liliaceae (hyacinth, tulip, garlic, onions)
- Primulaceae (primroses)
- Rutaceae (citrus fruits)
- Euphorbiaceae (castor beans, poinsettia, spurges)

Pollen grains of many species may cause hay fever, and this reaction is often referred to as "seasonal allergies." Inhalation of the pollen grains causes irritation of the nasal mucous membranes and can be effectively treated with antihistaminic drugs. Only light, wind-borne pollens cause hay fever and asthma attacks, but some fungal spores can also trigger a similar immune response. In early spring, tree pollens are the most common allergens (e.g., box elder, poplar, elm, birch, oak, hickory, and ash). In late spring and early summer, grasses are the most likely culprits (e.g., timothy grass, Bermuda grass, Kentucky bluegrass, redtop). In the late summer months of August and September, ragweeds (*Ambrosia* species) and saltbush (*Atriplex* species) are causative agents of hay fever.

Plants produce many toxic substances, and some common compounds can be classified as alkaloids, nitrates, glycosides, oxalates, resins, and toxic amines. Alkaloids from plants are well known, mostly because some of them have psychoactive and medicinal properties. Over five thousand alkaloids have been biochemically characterized, and many are bitter, solvent-soluble salts. Alkaloids may cause severe liver damage and fetal damage, and may have significant effects on the central nervous system (e.g., stomach pain, vomiting, and disorientation). Plants that contain alkaloid toxins in their seeds, flowers, leaves, or bulbs include the following:

- Wisteria
- *Galanthus* (snowdrops)
- Clivia
- Crinum
- Amaryllis
- Narcissus
- Yews
- Colchicum
- Gloriosa

Nitrates are a common poisonous substance for livestock raised on pasturelands. Nitrate-based fertilizers may be heavily applied on farmland and grazing pastures. These nitrates are taken up by the roots of some plants like *Amaranthus* (pigweeds), *Ambrosia* (ragweeds), and *Datura* (Jimson weed). Both animals and humans that eat wild edible plants such as corn, sorghum, oat hay, and some species in the nightshade, composite, mustard, and goosefoot plant families should be aware of the potential for nitrate poisoning. These plants can accumulate exceedingly high and toxic levels within their leaves and roots.

Toxic amines (proteins) are few in number in plant species but can have severe effects. Some plants that contain toxic amines include

- *Microcystis* (blue-green algae)
- *Claviceps* (ergot)
- *Phoradendron* (mistletoe)
- *Blighia* (akee)
- *Lathyrus* (wild and sweet peas)

Glycosides are sugars bonded to aglycones and are widely distributed among members of the plant kingdom. Cyanogenic glycosides are especially dangerous and are found in many common edible plants. Cyanogenic glycosides (e.g., amygdalin) are often found in the seeds of common fruits, including apples, peaches, plums, apricots, cherries, and almonds. Certain cassava or tapioca plants when eaten fresh can be quite toxic but after substantial processing can be safely ingested. Cardiac glycosides are also a common poison, especially after accidental ingestion by children. The heart stimulant digitalis is extracted from foxglove (*Digitalis*), which is a common garden flower. Other plants containing digitalis include

- *Nerium* (oleander)
- *Apocynum* (dogbane)
- *Scilla* (squill)
- *Convallaria* (lily of the valley)

Oxalates are found as acids or salts in many plants and give foliage a sour taste. Poisoning occurs when oxalate combines with blood calcium to form insoluble crystals that are painful and that can cause kidney damage. Soluble, toxic oxalates can be found in rhubarb leaves, green sorrel, and the berries of Virginia creeper. Many Arum family members contain calcium oxalate crystals and should be avoided as food items. These include the following:

- Jack-in-the pulpit (*Arisaema triphyllum*)
- Dragon root (*Arisaema dracontium*)
- Water arum (*Calla palustris*)
- Skunk cabbages (*Lysichiton americanum, Symplocarpus foetidus*)
- Dumbcane (*Dieffenbachia* species)
- Calla lily (*Zantedeschia aethiopica*)
- Elephant's ear (*Colocasia* species)

Resins are semisolid, and they burn easily and do not dissolve in water. Toxic resins are responsible for many poisonings and agonizing deaths that dramatically affect the central nervous system. Some plants that contain resins that are acutely toxic include

- Water hemlock (*Cicuta*)
- Milkweeds (*Asclepias*)
- Labrador tea (*Ledum*)
- Lords-and-ladies (*Arum maculatum*)

2.8 Historical Case Report: Poisoning by *Cicuta maculatum*

Water hemlock (*Cicuta maculatum*) is ubiquitous in geographic distribution throughout the U.S. and contains a dangerous convulsant called cicutoxin (a resin). Water hemlock is a member of the carrot and parsley family (Apiaceae) and has clusters of fleshy roots that smell like parsnips and are often mistaken as a wild edible. In fact, poisoning by water hemlock ranks among the top three causes of plant poisoning within the U.S. For children, the hollow stems may be appealing because they can be used as whistles; however, that can result in intoxication/poisoning. An adult person can die from ingesting as little as a single mouthful of the root or stem of the plant.

Death by convulsion is one of the most violent and painful manners of demise. Poisoning symptoms begin from 15 to 60 minutes after ingestion and proceed through nausea, tremors, and alternating periods of intense, painful convulsions followed by quiescent periods. Death typically results from exhaustion or suffocation. In ancient Greece, the philosopher Socrates chose death by water hemlock poisoning as his manner of execution. In 399 B.C., this was considered by his countrymen as a "humane" manner of execution.

2.9 Historical Case Report: A Botanical Mystery — the Great Hedge of India

Briefly mentioned in an obscure book of memoirs by a nineteenth-century British civil servant, the great hedge of India was planted in the 1850s as a physical barrier by the East India Company to aid in collection of a tax on salt. The hedge was mammoth — it stretched 1500 miles as part of the customs border that divided India from the Himalayas to Orissa. It was secured by approximately 12,000 men and consisted of many different plant species. In areas where a hedge could not be sustained, brief stone walls were constructed to maintain the continuity of the border. Some portions of the hedge were dry and composed primarily of dwarf Indian plum, or "jhurberry." The ideal hedge, however, was a living hedge that grew 10 to 14 feet high and 6 to 12 feet thick. The plant species intermixed to form a living hedge were thorny trees and shrubs and included the following species: Indian plum (*Zizyphues jujuba*), carounda (*Carissa curonda*), prickly pear (*Optunia* species), thuer (*Euphorbia* species), thorny creeper (*Guilandina bondue*), and babool (*Accacia catecha*). The existence of the hedge became somewhat of a mystery as it simply faded out of existence and people's memories so that one hundred years later, few could even recall if it had existed.

General References

Arora, D., *Mushrooms Demystified,* Ten Speed Press, Berkeley, CA, 1986.

Brown, L., *Grasses — An Identification Guide,* Houghton Mifflin Co., New York, 1979.

Gaillard Y. and Pepin G., Poisoning by plant material: review of human cases and analytical determination of main toxins by high-performance liquid chromatography–(tandem) mass spectrometry, *J. Chromatogr. B. Biomed. Sci. Appl.* 733, 181, 1999.

LeGrice, E.B., *Rose Growing Complete,* Latimer Trend & Co., Plymouth, England, 1976.

Moore, A.C., *The Grasses: Earth's Green Wealth,* Macmillan, New York, 1960.

Moxham, R., *The Great Hedge of India,* Carroll & Graf Publishers, Inc., New York, 2001.

Reich, L., *Uncommon Fruits Worthy of Attention—A Gardener's Guide*, Addison-Wesley, New York, 1991.

Richardson, J., *Wild Edible Plants of New England: A Field Guide*, Globe Pequot Press, Chester, CT, 1981.

Stewart, M.J. et al., Findings in fatal cases of poisoning attributed to traditional remedies in South Africa, *Forensic Sci. Int.* 101, 177, 1991.

Plant Cell Structure and Function

HEATHER MILLER COYLE

Contents

0-8493-1529-8/05/$0.00+$1.50

3.1 Introduction to Plant Cell Components

Essential to the identification and classification of a plant for recognition of value as evidence, one must know the parts of a plant and features that can aid in identification. This chapter focuses on the internal composition of a plant cell and how plant cells are organized into organs to serve specific functions essential to plant life. Plants and animals do not appear similar when comparing whole organisms; however, at the cellular level they are remarkably alike. The commonality of animal and plant cells argues for a common ancient ancestral cell type that has, over the course of evolutionary time, evolved into the more specialized animal and plant cells that we recognize.

The cellular structure of a plant cell as observed under light microscopy contains a cell wall, a nucleus, a plasma membrane, a central vacuole, cytoplasm, plastids and specialized plastid types, and often starch grains and crystals (see Figure 3.1). Cellular organelles can be observed easily with the use of inexpensive stains and light microscopy, or with transmission and scanning electron microscope techniques.

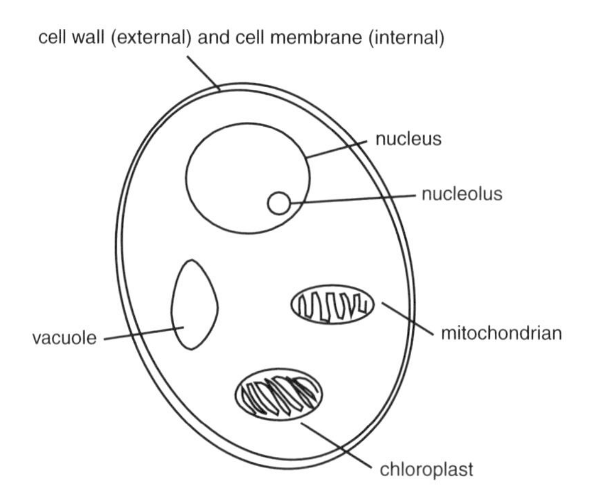

Figure 3.1 A schematic representation of a single plant cell is presented here. Plant cells consist of their component parts, including the cell wall, cell membrane, nucleus, mitochondria, chloroplasts, ribosomes, and other important cellular substructures.

3.1.1 The Cell Wall

Cell walls provide support for the plant and offer protection to the cell. Cellulose fibers and jelly-like pectin comprise the primary plant cell wall. Cell walls do not isolate cells entirely from one another. Some exchange between cells can occur through specialized channels called plasmodesmata that allow for cytoplasmic continuity between cells. Cell walls naturally inhibit growth and cell division; therefore, cells that are designed for continued growth have relatively thin plant cell walls. Some plant cells have more customized cell walls that become lignified and reinforced to assist in transport, protective, or support functions. These specialized cell walls are called secondary cell walls.

It is possible to "spectrotype" cell walls using Fourier transform infrared spectroscopy (FTIR). Mutations in genes that encode the cellular machinery involved in the synthesis, assembly, and disassembly of the cell wall have provided plants with defined alterations in wall composition and architecture. Cell wall biogenesis can be affected at the following stages:

- Substrate synthesis
- Polymer synthesis
- Secretion of Golgi-derived materials to the cell surface
- Assembly of polymers into a characteristic architecture
- Architectural remodeling during growth and differentiation
- Turnover and recycling of materials after their function has been completed

The discriminate analysis of FTIR as a robust method to identify cell walls in different plant species has been developed based on analysis of cell wall mutations in model systems of maize and arabidopsis. Mutants for which a specific defect has been deduced are valuable resources to better define the spectral deviations observed by FTIR between cell walls and to correlate these deviations with thc chcmical and physical alterations in architecture and composition responsible for them. For example, mutations in cellulose syntheses provide distinctive spectra that are reduced in wave numbers characteristic of cellulose. When these infrared mutant "spectrotypes" are compared with those of standard polysaccharides and artificial polysaccharide matrices, they comprise a rudimentary library to characterize additional mutants. A comparable library could be useful for characterizing different plant species.

3.1.2 The Nucleus

The cell nucleus is oval and contains deoxyribonucleic acid (DNA) molecules that are the heritable information for regulating cell activity. The nuclear contents are separated from the rest of the cell by the nuclear envelope. Within the nuclear envelope is the nucleoplasm, which is the gel-like substance that houses the chromosomes (DNA-protein complexes). Eukaryotic plant cells have DNA and ribonucleic acid (RNA) molecules within a nucleus; prokaryotic cells lack a membrane-bound nucleus and their DNA and RNA is free-floating within the cytoplasm. Within plant cell nuclei, often multiple nucleoli are present. These nucleoli are comprised of RNA and proteins and are precursors of ribosomes. Ribosomes are often attached to the outer surface of the outer nuclear membrane.

3.1.3 Ribosomes

Ribosomes are small particles that occur in many places throughout the cell, including the cytoplasm, on the outside of the endoplasmic reticulum membranes, and in the nucleus, the mitochondria, and the chloroplasts. Ribosomes are the sites for protein synthesis and consist primarily of RNA and histone protein. They function for the assembly of amino acids into polypeptides.

3.1.4 The Plasma Membrane

The plasma membrane is a substance forming the outer boundary of the plant cell and lies internal to the cell wall. The cell membrane plays an important role in regulating the chemical composition of the cell. Other membranous structures occur in the cytoplasm, including the vacuole, the spherosome (lipid bodies), microbodies (sites for glycollic acid oxidation), the Golgi apparatus (involved in secretion), and the nucleus. In addition, the endoplasmic reticulum (ER), a complex system of two unit membranes with a narrow space between them, is thought to be involved in intracellular transport and perhaps wall formation and some types of secretions (e.g., nectar).

3.1.5 The Vacuole

A vacuole is a watery compartment surrounded by a membrane called the tonoplast. The vacuole could contain many different substances, including, but not limited to, tannins, protein bodies, sugars, organic acids, flavenoid pigments, calcium oxalate, and phosphatides. Vacuoles serve a function in regulating water and solute content, and in such processes as osmoregulation and digestion, as well as for storage.

3.1.6 The Cytoplasm

The cytoplasm is fluid and can move (cytoplasmic streaming); however, its gel-like structure can serve to support other organelles located outside the nucleus. The major component of the cytoplasm is water (85 to 90%), and it is a viscous substance that is transparent in visible light. The cytoplasm is contained within the plasma membrane but is delimited further from other organelle-bound structures like vacuoles.

3.1.7 The Plastids

Numerous types of plastids exist: chloroplasts, leucoplasts, and chromoplasts. All plastids originate from proplastids located in egg cells and young, dividing (meristematic) cells. Chloroplasts are the sites of photosynthesis for conversion of light energy into carbohydrates. Chloroplasts are green due to the chlorophyll pigment contained within them. Leucoplasts store starch (amyloplasts), fats (elaioplasts), and proteins (proteinoplasts), and are typically found in tissues not exposed to light (e.g., roots). Chromoplasts are found in petals, fruits, and some roots and store carotenoid pigments. Due to the coloration of the pigments, these plant tissues will appear yellow, red, or orange in color.

3.1.8 Ergastic Substances

Waste products and reserved molecules produced by cells include the following: starch, crystals, proteins, silica bodies, and tannins. Starch is a long-chain carbohydrate that appears as grains. Starch grains are formed in chloroplasts and are then broken down and reformed in amyloplasts. Starch from plants can have nutritional and economic value, and much of the starch we use in commercial products or as food items are from cereals (wheat, maize, and rice), potato, tapioca, sago, and arrowroot. Proteins can be amorphous or crystalline, and are common in the endosperm and embryo of seeds. Crystals are formed by the excess deposition of inorganic molecules, and they can take on a variety of forms: prisms (rectangular or pyramids), druses (spheroid aggregates), raphides (elongated with sharp tips), styloids (elongated with blades on end), and crystal sand (tiny spherical, in masses). Most crystals form within the vacuoles for storage, and their appearance can be helpful in the classification of different species.

3.1.9 Spherosomes

Spherosomes are small, round lipid bodies found within the cytoplasm that are diagnostically opaque after fixation with osmium tetroxide under scanning electron microscopy. Their role is somewhat unclear, but it has been speculated that these oil-filled vesicles may be generated from the endoplasmic reticulum and become detached and free-floating in the cytoplasm. Other oils and fats are present in the vacuole as reserve lipids and are very common in seeds and fruits. Waxes are mostly found on the surface of leaves, stems, and fruits and serve to protect the plant.

3.1.10 Microtubules

Microtubules are hollow, long, straight structures composed of protein subunits and are located in the periphery of the cytoplasm, closest to the cell wall. They are also found in the mitotic and meiotic spindles and in another cell division-related structure, the phragmoplast, which appears in telophase. Since microtubules are associated with cell division and near cell walls where growth is still occurring, one theory is that microtubules provide direction for developing microfibrils.

3.1.11 Mitochondria

Mitochondria are involved in cellular respiration and can be seen under a light microscope with the Janus Green B stain. They can look lobed, elongated, or spherical when cells are sectioned. These organelles have two unit-membranes: the internal membrane forms many folds into the matrix (primarily composed of protein), and the outer membrane surrounds the inner membranes. The mitochondria are involved in energy conversion and contain enzymes used for the Krebs cycle.

3.2 Basic Plant Tissues

3.2.1 Meristem

A plant embryo, in the early stages of development, will undergo continuous cell division until the adult plant body is formed. After plant cell differentiation into specialized tissues has occurred, embryonic-like cell division is restricted to certain regions of the plant called meristems (see Figure 3.2). The definition of meristematic tissue includes cells that have the ability to continue on with unlimited numbers of cell division. Other young tissues

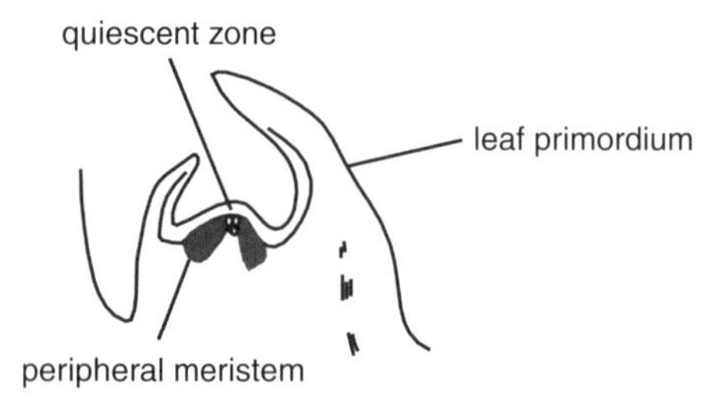

Figure 3.2 The meristematic region of a plant is defined as an area of actively dividing cells that can give rise to new organs and tissues. Apical meristems are located at the tips of shoots and roots; lateral meristems are cylinders of cells found on the sides of roots and shoots.

may continue to exhibit cell division, but those numbers are limited. Meristems are classified into different types based on the following:

- Position in plant body
- Their structure
- Their function
- Their stage of development
- Their origin
- The types of tissues they produce

Based on their position in the plant body, meristems can be apical (located in the apices of main and lateral shoots and roots), intercalary (located between mature tissues), or lateral (positioned parallel to the circumference of the organ where they are located). Meristems are often referred to as primary (derived from embryonic cells) or secondary (derived from mature differentiated tissues) based on the tissues they originate from. Meristems are responsible for generating new roots, shoots, and flowers on the body of the plant as well as extending the length of parts of a plant body, such as the stem. The cellular patterns of division within these different types of meristems have been mapped and studied extensively in order for developmental biologists to better understand the ordered plan for generating a plant body and specific plant organs.

3.2.2 Parenchyma

Parenchyma is comprised of living cells that may differ in shape and function. These cells retain their ability to divide even when mature and thus are able to assist in wound recovery and tissue regeneration. Parenchyma cells are considered somewhat primitive because many of the phylogenetically lower plants consist only of parenchyma cells. The following parts of a plant contain parenchyma cells:

- Pith (the ground tissue in the center of the stem and root)
- Root cortex (tissue between the vascular cylinder and root epidermis)
- Shoot cortex (tissue between the vascular cylinder and shoot epidermis)
- Pericycle (the ground tissue between the conducting tissues and the endodermis)
- Leaf mesophyll (the photosynthetic tissue between the two epidermis layers)
- Xylem (tissue that conducts water in vascular plants)
- Phloem (tissue that conducts nutrients in vascular plants)
- Fleshy fruits

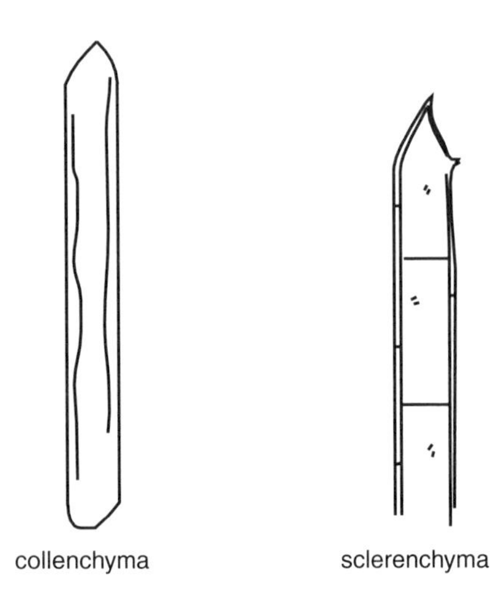

Figure 3.3 Collenchyma and schlerenchyma are types of plant tissues that are used for support of the plant body. Collenchyma cells usually lack a secondary wall and are found in young growing tissue; schlerenchyma cells have a secondary wall with strengthening lignin and are not able to elongate in growth.

3.2.3 Collenchyma and Sclerenchyma

The supporting tissues of a plant are called collenchyma and sclerenchyma (see Figure 3.3). Collenchyma is made up of living, elongated cells with typically uneven thickening of the cell wall. Immature collenchyma is stretchable and irreversibly expands as an organ grows and develops. Mature collenchyma is more brittle and less plastic and, like parenchyma, may contain chloroplasts and tannins. In plants exposed to shaking or heavy winds, collenchyma tissue will become increasingly thicker to aid in support of the plant body. Collenchyma is located in the following regions of a plant body:

- Stems (in longitudinal strips or as a complete cylinder)
- Leaves (on one or both sides of a vein; along margins of a blade)
- Floral parts
- Roots
- Fruits

Sclerenchyma also functions to support the plant body and may result from lignification and wall thickening of collenchyma tissue. Unlike collenchyma, however, schlerenchyma is elastic (which means it can expand and contract) rather than plastic (capable of expansion only). Sclerenchyma can be further subdivided into two types: fibers (long cells) and sclereids (short cells). Although size is a general trend for the classification of fibers and sclereids, it is not very accurate. A true classification is based on the origin of each sclerenchyma type as fibers develop from meristematic cells and sclereids originate from parenchyma cells that have thickened secondary walls. Fibers may be present as single cells or as groups throughout the plant body. Most often, they comprise vascular tissues such as xylem and phloem. Sclereids also can be found as single cells; however, most often they are present as masses of hard cells within soft parenchyma tissues. Interestingly, the shells of walnuts, for example, are composed entirely of sclereids.

3.2.4 Xylem

The vascular system of higher plants plays an important role for both physiology and phylogeny and has been used in classification of various plant species. The role of the xylem is to transport water and solutes from the roots to the arial portions of the plant body. The xylem is a complex tissue consisting of numerous cell types, including tracheary elements (tracheids and vessel members), fibers, sclereids, and parenchyma cells. The tracheary elements, in particular, are used for evolutionary phylogenetic study. The cell features of tracheary elements useful for this type of study include

- Length of the element
- Diameter of the element
- Thickness of the cell wall
- Shape of the element when viewed in cross section
- Type of pitting
- Type of perforation plate

3.2.5 Phloem

The phloem is a vascular conducting tissue for transport of photosynthetic products. The sieve elements are the cells responsible for carrying out this function. Other cell types found within phloem tissue include specialized parenchyma cells (companion cells, albuminous cells), fibers, sclereids, resin ducts, and lacticifers (cells containing latex). The position of phloem in a stem is usually external to that of the xylem tissue. However, in some plant families (e.g., Cucurbitaceae, Myrtaceae, Compositae, Solanaceae), the phloem can also be located inside the xylem. The identifying feature of a sieve element is the lack of a nucleus and a viscous substance called P-protein. P-protein stains easily with cytoplasmic stains, and with the use of an electron microscope, its features can be useful for distinguishing some plant species. P-protein appears present in all dicotyledonous species but seldom in monocots, gymnosperms, and pteridophytes. The presence of P-protein in dicot species sieve elements may be observed as many forms including filamentous, tubular, granular, and crystalline.

3.2.6 Secretory Cells

Plant cells commonly secrete a variety of substances, including waxes, forms of latex, and resins. The term *secretion*, in the context of this book, will refer to the active elimination of a substance from a cell. Resin ducts are quite common in conifers. In some species, they are present as a normal state; in others, they are the result of injury to the plant. In the genera *Pinus*, *Larix*, *Pseudotsuga*, and *Pinus*, resin ducts appear as typical features of wood, and the substance they produce is commonly called "pitch." Plant "gums" are the result of a process termed *gummosis*, which is the metamorphosis of primary and secondary cell wall matter into disorganized substances called gums. As the cell walls break down, the cavity is filled with starches and gums. Gummosis occurs often as a result of mechanical injury to the plant through the action of insects or diseases, or by physiological disturbance. Some gums are commercially important and can be distinguished from one another using capillary electrophoresis and polarized light microscopy. These types of chemical analyses are useful when determining purity of a compound, which may be important to food items or manufacturing processes. Latex occurs as a suspension or emulsion and is common to many angiosperms. The chemical composition and color of latex differs in various plant species and may be used as a classification feature. Within the latex, salts, organic acids, and other substances may occur as in the following:

- Compositae (sugars)
- *Musa* (tannins)
- *Papaver somniferum* (alkaloids)
- *Carica papaya* (enzymatic papain)

Perhaps the best example of a secretory plant, and one with worldwide economic importance, is the Para rubber tree (*Hevea brasiliensis*). The latex of *H. brasiliensis* contains approximately 30% rubber, and it is grown in Central America, the West Indies, Brazil, Liberia, Sumatra, Ceylon, Java, eastern India, and the Malayan Archipelago. During World War II, when rubber was inaccessible from Asia, other rubber-containing plants were identified as secondary economic sources of rubber:

- *Castilla* (Panama rubber)
- *Manihot* (Ceara rubber)
- *Parthenium argentatum*
- *Taraxacum kok-saghyz*
- *Hancornia*
- *Landolphia*
- *Cryptostegia*

Waxes are deposited commonly on the surface of the cuticle for protection of plant leaves and fruits. These waxes may form a continuous layer or may be deposited in granules, rods, scales, or platelets. The wax reduces the wettability of the plant leaf surface and lends a shine to increase the appeal of certain fruits.

3.2.7 Epidermis

The epidermis is the outermost cell layer located on the exterior of leaves, stems, flowers, fruits, seeds, and roots. Both morphologically and functionally, the epidermis is not a uniform tissue as it is comprised of numerous different cell types, some of which have discrete functions. The epidermis persists throughout organs until secondary thickening of cell walls occurs. In some plants, the epidermis may be replaced by cork tissue as they age, especially if they have a long life span. The ordinary cells of the epidermis may vary in size and shape, but they are all tightly compacted without intercellular spaces. The epidermal cell walls can also vary in thickness, and on the outer surface, pectin may be found. Cutin, a fatty substance, stains red with the cellular stain Sudan IV and may be embedded within or along the surface of the outer walls of the epidermis. This layer of cutin forms a structure called the cuticle, which may be complex and diagnostic for certain plant species. The cuticle surface may be smooth, ridged, or furrowed. From early geological studies, recovered cuticles have retained the shape and orientation of the mesophyll cells below, providing important information to classify these plant forms. Within the epidermal layer are also stomata, which are intercellular spaces surrounded by specialized guard cells that control the opening and closing of these spaces. Stoma (the plural of stomata) are typically present on the arial portions of rhizomes, leaves, and stems as they play an important role in regulating transpiration rates in plants. In addition, the single or multicellular appendages of the epidermis are called trichomes ("plant hairs"). The use of stomatal spacing and trichome characteristics in phylogenetic classification is well known in forensic botany.

3.2.8 Periderm

The periderm is a secondary plant tissue consisting of three parts: phellogen (cork cambium), phellem (cork), and phelloderm (parenchyma-like tissue). The cork functions as a protective layer to protect the plant as the epidermis dies and is shed. Cork is present in the roots and stems of many dicotyledonous plants and, with all of the parts together, is commonly known as "bark." The type of bark present on trees, for instance, can be used in many cases as a taxonomic classification character (e.g., rough, smooth, and continuous). Some genera have overlapping regions of periderm that produce a scaly bark (e.g., *Pinus*, *Pyrus*). Other genera (e.g., *Vitis*, *Clematis*, *Lonicera*) form continuous cylinders of periderm called "ring bark." Still other plant species are intermediate between the scaly and ring barks (e.g., *Arbutus*, *Platanus*, and *Eucalyptus*).

Commercial cork is produced from many trees, in particular, *Quercus suber*. Cork is valued because it is impenetrable to gas and liquid and has strength and elasticity and is lightweight. Stripping the periderm from a 20-year-old tree is the first step in forming commercial cork. The exposed cells of the phelloderm and cork die, and a new phellogen is rapidly formed as a wound response. Within 10 years, this second phellogen is ready for harvest. This process continues at 10-year intervals until the tree is approximately 150 years old. The dark brown spots that may be noted within commercial cork (e.g., wine bottle corks) are lenticels, isolated areas in the periderm consisting of suberized and nonsuberized cells.

3.3 Common Staining Techniques and Laboratory Exercises

Cell shape and orientation of certain structures within a cell can be helpful in classification of a species. In order to learn about plant anatomy and specific plant structures within the plant body plan, it is important to take a practical approach. In the following sections we describe examples of inexpensive laboratory exercises that can be performed as training for forensic botanists.

3.3.1 Epidermis

This is a surface tissue of young plants and herbaceous plants that lack lateral meristems. The epidermis may only be one cell layer in thickness.

1. Obtain a leaf from *Tradescantia* (Spiderwort) or *Zebrina* (Wandering Jew).
2. Tear the leaf at a 90° angle to the veins. Look for a small region of the lower (purple) epidermis at the edge of the tear.
3. Using forceps and a clean razor blade, make a wet mount of this purple tissue with a drop of water on a microscope slide. Place a coverslip on top of your sample. A drop of methylene blue may be added to increase the contrast between the cell types.
4. First examine the slide under low magnification (4X or 10X). Two types of cells should be present: epidermal cells and guard cells. Guard cells occur in pairs and surround the stomata. The epidermal cells should form an interlocking layer or sheet of cells with the guard cells interspersed in pairs throughout the layer.

3.3.2 Trichomes

Specialized epidermal cells (trichomes) appear as "hairs" on the surface of the plant epidermis. They can be one cell or multicellular, and are useful features for species classification. Some trichomes reflect light as an optical illusion for insects or contain stinging chemicals for protection from predators.

1. Examine leaves of a *Coleus* plant and a tomato (*Lycopersicon*) plant under a dissecting microscope. Locate the trichomes. Carefully record their structure, noting the number and shape of the cells as well as the spacing on the leaf surface.

3.3.3 Periderm

In 1665, Robert Hooke described the cellular nature of cork, which comprises the most abundant part of the periderm in woody plants.

1. Prepare a dry mount of a very thin section of cork tissue. Cork samples may be obtained from supermarket wine bottles. The section does not have to be very large, in fact, smaller than a millimeter is sufficient. Do not add water or a coverslip.
2. Observe the tissue with both low (4X) and then high (100X) magnification. Note the uniform size and general appearance of the cork cells. These are dead cells with no cytoplasm. What are observed under the microscope are components of a waxy cell wall.

3.3.4 Collenchyma

Collenchyma consists of living elongated cells that have an uneven thickness to their primary cell walls. They are usually arranged in strands in stems and petioles. Collenchyma cells are classified as either "angular" or "lamellar" by their patterns of cell-wall thickening. Angular collenchyma has cell walls in which the corners are extra thick (e.g., celery). Lamellar collenchyma has thickenings on both the inner and outer cell walls (e.g., tomato and sunflower).

1. Cut a thin section of a celery petiole and place on a microscope slide without a coverslip. Observe under the dissecting microscope. What different cell types are distinguishable?
2. Stain the section with toulidine blue for one minute. Rinse the excess stain and observe again under the dissecting microscope. Describe what you see.
3. Under low magnification (10X), locate the small clusters of cells on each of the ridges along the outer portion of the petiole. These are the strands of collenchyma tissue (e.g., strings). Do not misinterpret collenchyma for the vascular bundles located in the interior of the section.

4. Peel a strand of collenchyma from the petiole. How much tension is required to break the strand?

3.3.5 Sclerenchyma

Sclerenchyma serves as structural support of stems, leaves, and other plant parts that are no longer elongating. Sclerenchyma cells are generally dead at maturity, and there are two common types of sclerenchyma cells: sclereids and fibers.

1. Prepare a wet mount of a small amount of the fleshy part of a pear fruit. The small, slightly hard cells that you notice when you eat a pear are clusters of sclereids (groups of a few cells or perhaps 30 cells).
2. Prepare a second wet mount of pear tissue and stain for one minute with phloroglucinol, a highly acidic stain that reacts with lignin. Observe the stained specimen for sclereid cells again. What is their general shape and thickness?

3.4 Summary

The accurate forensic identification of different plant species is often dependent on knowledge of plant anatomy and different plant cell types. Botanical evidence may be significant, as in a large wood chip that is readily identifiable by wood grain and texture. However, fragments of plant material may be extremely small and may require thorough microscopic examination to identify, first, what plant part it is. Second, species identification may be possible if an appropriate reference guide is available for comparison against the evidentiary sample. Simple staining procedures and a basic recognition of cellular patterns could be the key step for solving the case. In addition, further forensic research studies can be performed to aid in establishing good pictorial reference guides for local plant material for the different quadrants of the U.S.

General References

Bruce, D.M., Mathematical modelling of the cellular mechanics of plants, *Philos. Trans. R. Soc. Lond. B. Biol. Sci.*, 358, 1437, 2003.

Fahn, A.H., *Plant Anatomy*, 3rd edition, Pergamon Press, New York, 1982.

Flurer, C.L., Crowe, J.B., Wolnik, K.A., Detection of adulteration of locust bean gum with guar gum by capillary electrophoresis and polarized light microscopy, *Food Addit. Contam.*, 17, 3, 2000.

Friedman, W.E., Cook, M.E., The origin and early evolution of tracheids in vascular plants: integration of palaeobotanical and neobotanical data, *Philos. Trans. R. Soc. Lond. B. Biol. Sci.*, 355, 857, 2000.

Genard, M., Fishman, S., Vercambre, G., Huguet, J.G., Bussi, C., Besset, J., Habib, R., A biophysical analysis of stem and root diameter variations in woody plants, *Plant Physiol.*, 126, 188, 2001.

Jagels, R., Visscher, G.E., Lucas, J., Goodell, B., Palaeo-adaptive properties of the xylem of Metasequoia: mechanical/hydraulic compromises, *Ann. Bot.*, 92, 79, 2003.

Niklas, K.J., The influence of gravity and wind on land plant evolution, *Rev. Palaeobot. Palynol.*, 102, 1, 1998.

Modes of Plant Reproduction

4

HEATHER MILLER COYLE

Contents

0-8493-1529-8/05/$0.00+$1.50

4.1 Introduction

Life is derived from previously existing life. The method or mechanism for how life is derived can be used as a classification scheme. Each living organism gives rise to generally more of its kind. The mechanism for how organisms reproduce can vary significantly. In plants, one form of reproduction can be favored over another, but within a plant population multiple modes of reproduction may exist. These different mechanisms for reproduction or procreation are important to determine when considering generating a DNA profile from a botanical sample. The genetic principles behind the reproductive mechanism can affect the overall interpretation of how often one might expect a particular genetic profile to be encountered within a population.

4.2 Forms of Plant Propagation

4.2.1 Sexual Reproduction

The process of reproduction involving the fusion of haploid (1N) gametes to form a diploid (2N) zygote is called fertilization. The zygote develops into a new individual organism. The life cycle of most plants is characterized as the alternation of haploid and diploid generations in which each generation takes turns producing the other. The diploid plant (2N, sporophyte) produces a gametophyte (1N, seed, or pollen). Gamete-producing plants are called gametophytes; spore-producing plants are called sporophytes. The sexual process in seed-producing plants involves fertilization within the protective female tissues analogous to mammalian reproduction. After fertilization (syngamy), a developmental sequence results in the development of a seed. This is an embryonic plant, surrounded by endosperm (food for maintenance and growth) and a seed coat (for water conservation and protection). Seed production occurs in cones (e.g., conifers) or in flowers (e.g., angiosperms). Plants are annuals (which means they grow and produce seed within the same calendar year), biennials (they grow vegetatively in the first year, and then produce seed in the second year), or perennials (they live indefinitely and flower each year). Seeds are produced through the process of meiosis, which results in the recombination of plant genomes and plant progeny that share 50% of each of the parents' DNA. Therefore, individual seeds have DNA profiles that are different from each of their parents. However, due to the percentage of shared DNA based on the rules of Mendelian inheritance, paternity and maternity testing can be performed on seed-generated plant progeny just as with human DNA testing. (See Figure 4.1.)

Flower arrangements may be different depending on the plant species. Monoecious species have both male and female flowers on the same plant.

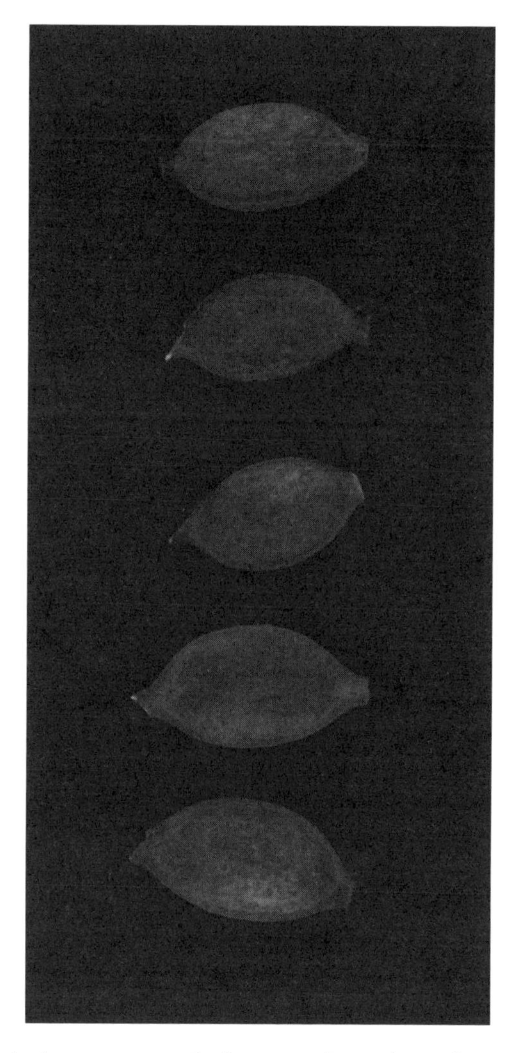

Figure 4.1 Seeds from trees and plants such as the Palo verde seeds shown here have their own genetic composition through the mechanism of sexual reproduction.

Dioecious species have female flowers on one plant and male flowers on the second plant (e.g., ginkgo, marijuana). Monoecious species can self-pollinate (pollen from the same plant fertilizes the flower on that plant) or cross-pollinate (pollen from one plant fertilizes flowers on a separate plant). Hermaphroditic plants are possible; this term is used most often when the typical situation is dioecy, but an exceptional plant will show monoecy (e.g., asparagus). Obligate self-pollination refers to pollen release within a flower before it fully opens; therefore, it cannot receive pollen from an outside source (e.g., some tomato cultivars). Monocotyledonous species typically have floral parts in threes or multiples of threes. Dicotyledonous plants have floral parts that group typically in multiples of fours and fives.

Flowers are typically grouped into wind-pollinated plants and animal-pollinated plants. Wind-pollinated plants have smooth, dry, and light pollen. Their floral stigmas are long and feathery, and the flowers themselves are small and inconspicuous. Wind-pollinated flowers generally yield a single fruit per flower. Pollen from animal-pollinated plants is heavy and adherent. The floral stigmas are sturdy, rounded, and compact. Typically the flowers are showy and scented, and produce nectar. Many petals have ultraviolet nectar guides that are visible to the insect eye. Animal-pollinated flowers generally yield fruits with multiple seeds.

4.2.2 Asexual Reproduction

Asexual reproduction is defined as the process by which organisms multiply without the formation or fusion of gametes (special sex cells). Asexual plant reproduction is also called vegetative propagation, an extension of the ability for plants to have indeterminate growth. Fragmentation is the separation of a parent plant from parts that can reform into whole, intact plants, resulting in genetic clones. The ability for some plants to form multiple generations of identical clones is a substantial departure from the genetic system most frequently observed in humans. Although identical twins can result from human sexual reproduction, large populations of identical clones are not present in human populations. In addition, the ability to propagate plant cuttings in tissue culture or to form hybrid genetic offspring (genetic mosaics) through grafting is common in some plant species but rare in other organisms.

4.2.2.1 Cloning

Cloning is defined as either of the following:

- Any of two or more genetically identical individuals produced from one parent by asexual reproduction
- Either of the two individuals produced by the splitting of a young embryo

Leaves and petioles (cuttings) removed from a "mother" plant and treated with plant hormones to induce root formation result in genetically identical plant clones. This is a common form of plant propagation for species with high economic value that need to maintain a continuous phenotype. In particular, high-tech illicit marijuana propagation favors this form of plant reproduction since it maintains the high potency THC of the parent plant in all of the daughter progeny. Clones may be grown hydroponically (with liquid nutrients) or, after root formation, planted in soil for continued growth and floral bud development.

Plants that can be effectively propagated as "cuttings" include

- *Buddelia* (butterfly bush)
- *Catalpa* (Indian bean)
- *Cornus* (dogwood)
- *Ginkgo* (maidenhair tree)
- *Syringa* (lilac)
- *Vaccinium* (blueberries)
- *Taxus* (yews)
- *Buxus* (boxwood)
- *Choisya* (Mexican orange)
- *Ilex* (holly)
- *Arbutus* (strawberry tree)
- *Forsythia* (forsythia)
- *Lonicera* (honeysuckle)
- *Rosa* (roses)
- *Salix* (willow)
- *Vitis* (some grapes)

4.2.2.2 Grafting

A graft is defined as the transfer of a small portion of one organism to a relatively large part of a second organism, which is later transplanted. In plants, this is typically called a heterograft if one plant species is grafted to another species. An autograft involves the transfer of an individual plant to another plant of the same species. Grafts are often used in horticulture where the plant to be cultivated (scion) is grafted onto the rooted portion (rootstock). The purpose of grafting is to combine the features of both plants (e.g., grafting a disease-susceptible ornamental rose scion to a disease-resistant rose rootstock). Grafting is exceedingly common for floral ornamentals and fruits. The genetic DNA profile will be different for the scion and rootstock, respectively.

4.2.2.3 Rhizomes

A rhizome is a horizontal underground stem that has leaves and buds and so is capable of vegetative propagation. The rhizome serves as a storage organ and is found among many flowering plants, such as *Iris*. Plants that are connected by these underground stems are genetically identical (e.g., clonal).

4.2.2.4 Stolons

A stolon, often called a runner, is a long, slender stem that travels along the surface of the ground. A stolon arises from the axil of a leaf and allows for rapid vegetative propagation. Along the runner are small buds from which roots may emerge and grow into the soil. The stolon will eventually wither away, leaving multiple daughter clones. Stolons are common in plants such as strawberries and creeping buttercup. Like rhizomes, plants connected or produced by common stolons will have identical DNA profiles.

4.2.2.5 *Apomixis*

Apomixis is the development of a seed or embryo in plants without fertilization or meiosis. This pseudosexual reproduction is common in garlic and citrus species, and the genetic constitution of these seeds needs to be carefully considered if one is trying to generate a DNA profile. Apomixis is a naturally occurring process, and the evolutionary value of apomixis is unclear.

4.2.2.6 *Plant Cell and Tissue Culture*

Along with traditional plant breeding methods, modern manipulation of plant reproduction through the use of new biotechnology is becoming more widely utilized. Meristematic regions of plants can be induced with proper hormone treatment to reproduce *in vitro* and can be propagated in petri plates and test tubes into intact plants. These plants should be clonal; however, somatic mutation (somaclonal variation) may also occur at a greater rate in some plant species that are reproduced through the tissue culture process. For example, Gerbera daisies are propagated at tissue culture facilities (such as Weyerhauser Inc., in Apopka, FL) to ensure the integrity of flower uniformity and color for landscaping purposes.

4.3 Genetic Modification

Genetic modification of food plants is a controversial topic. Some feel that gene manipulation creates some global hazards that bear consideration. Others feel that genetic engineering is simply a method for the more rapid alteration of a plant genome and is no different than traditional plant breeding methods. Currently, approximately 50% of soybeans and 33% of cotton crops in the U.S. are genetically modified. In addition, canola, squash, and potatoes are genetically modified and have been on the U.S. market for approximately four years. The National Academy of Sciences has stated that no strict distinction between health and environmental risks posed by genetic engineering exists when compared to traditional breeding practices. "Golden rice" contains three transgenes: two from daffodil and one from a bacterium. These transgenes increase beta-carotene production and other carotenoids. In the developing world, dietary deficiencies in carotenoids and vitamin A lead to blindness and cause one to two million early childhood deaths. Other goals for plant genetic engineering include field-testing of genetically modified alfalfa containing the nontoxic beta chain of the cholera toxin in the hopes of developing an edible vaccine. With all of the positive potential use of genetically modified plants, the possibility of poisoning other organisms or the unintentional outbreeding of a transgenic crop with wild plant relatives needs to be considered. In a well-publicized study, researchers showed that pollen that had been altered to produce *Bacillus thuringiensis* toxin could reduce growth rates and often kill monarch caterpillars when pollen was fed to them along with their milkweed diet.

4.4 Artificial Seeds

Artificial seeds are embryos encapsulated in synthetic gelatin capsules. They are more expensive to purchase than traditional seeds, but the benefits are that plants grow more uniformly and are easier to harvest. There are numerous companies that offer artificial seeds, many of which are available over the Internet.

4.5 Plant Genomics

The field of plant genomics is essentially the study of plant genetics as applied to the whole genome. Plant genomics may include DNA sequencing of model plant species to identify genes (coding regions of the DNA) and functional genomics (determination of the function of a protein product in the context of cell differentiation and development). In 2003, the National Science Foundation (NSF) added $100 million to ongoing funding of research projects for extending the collaboration of researchers to form virtual centers for the study of plant genomics. Perhaps the best-known model system is *Arabidopsis thaliana* (thale cress), a member of the mustard family. This plant was selected for genome sequencing because it has five chromosomes and grows quickly, yielding up to eight new generations in a single year. The 118.7 million base pairs encode approximately 25,000 genes, of which 100 genes bear some resemblance to those found in humans. Identification of the genes and an understanding of how this model plant system is regulated may eventually aid in the development of greater plant yields and the ability to boost nutritional values in crop species. In addition, a more comprehensive understanding of the complex genetic basis for flower and pollen development may allow for future genetic modification of reproductive strategies for economically valued horticultural and crop plant species.

4.6 Databases as Repositories for DNA Information

As more and more genes are mapped and sequenced from different species, there is a need for a central repository for the biological information that is being generated from these studies. These databases result from large cooperative projects to organize genetic information in a convenient manner for public access to researchers.

4.6.1 GenBank

GenBank is a U.S. government-sponsored repository of protein and nucleic acid sequences that is maintained by the Los Alamos National Laboratory. GenBank has criteria for the acceptance of new data into the database, and each new entry is given an accession number for ease of identification and tracking. All scientific publications are required to reference these accession numbers for clarity and to avoid confusion between overlapping but nonconcordant genetic nomenclature. Once data have been submitted to GenBank, they are publicly accessible through a website interface. This website also contains useful references and links to PubMed (medical and scientific literature references). The DNA listings in GenBank are organized into subdivisions based on taxonomic categories. More general categories include structural ribonucleic acids (RNAs), synthetic deoxyribonucleic acids (DNAs, oligonucleotides), and newly arrived sequences that have not been classified into a category yet.

4.6.2 EMBL Databank

This databank is similar to GenBank but is maintained by the European Molecular Biology Laboratory. GenBank and EMBL exchange data freely between their databanks.

4.6.3 PIR and SWISS-PROT Databanks

PIR (Protein Identification Resource) and SWISS-PROT are repositories for protein amino acid sequences. PIR is a U.S. government-funded program and is maintained by the national Biomedical Research Foundation. SWISS-PROT is the European counterpart to PIR.

4.6.4 DDBJ Databank

The National Institute of Genetics in Mishima, Japan, maintains the DNA Data Bank of Japan (DDBJ). This databank is the Japanese equivalent to GenBank.

4.6.5 Plant DNA C-Values Database

The DNA amount in the unreplicated gametic nucleus of an organism is referred to as its C-value, irrespective of the ploidy level of the taxon. The Plant DNA C-values Database currently contains data for 3927 different Embryophyte plant species. It combines data from the Angiosperm DNA

C-values Database (release 4.0, January 2003), the Pteridophyte DNA C-values Database (release 2.0, January 2003) together with the Gymnosperm DNA C-values Database (release 2.0, January 2003), and the Bryophyte DNA C-values Database (release 1.1, January 2003). This database is maintained by the Royal Kew Gardens.

4.6.6 TIGR (The Institute for Genomic Research)

As part of an Arabidopsis Genome Program, TIGR has generated sequences from several thousand random clones from the Landsberg erecta accession of *Arabidopsis thaliana.* These sequences can be used to generate "electronic polymorphisms" that can be used to generate new, targeted markers for use in genetic mapping studies.

As part of a National Science Foundation-funded consortium, TIGR has generated more than 60,000 expressed sequence tags from various potato cDNA libraries. These sequences are single-pass sequences from the 5' end of cDNA clones. A total of 10,000 good sequences have been generated from 5 core potato tissues (stolons, tubers, leaves, shoots, and roots). A total of 5000 good sequences have been generated from a library constructed from leaf tissue challenged with an incompatible strain of *Phytophthora infestans.* Approximately 15,000 cDNA clones were resequenced as part of the microarray component of the project, and both 5' and 3' sequences are available for these clones.

4.7 Extraction of Plant DNA from Dried Plant Rhizomes

As a first step in performing genomic research or genetic manipulation using molecular biology methods, the DNA from the cell nucleus needs to be extracted and separated from the other cellular components. There are many ways to extract DNA from plant material, and depending on the tissue source and the preservation conditions (e.g., fresh, dried, leaf, seed), some procedures may be more effective in purifying high quality and high yields of genomic DNA. In general, all plant DNA procedures use the following steps:

- Disruption of plant cells (mechanical grinding in liquid nitrogen or chemical disruption with detergents such as sodium dodecyl sulfate [SDS])
- Separation of DNA from other cell constituents (chemical methods, centrifugation methods, differential binding to solid supports or beads)
- Purification of DNA (precipitation with salts and alcohol or with polyethylene glycol [PEG])
- Quantification of DNA (spectrophotometry or visualization by staining)

Many commercial DNA extraction kits are available for purchase; however, some of the standard plant DNA extraction procedures are also interspersed throughout this book. The following protocol is especially useful for dried plant material such as might be found after years of storage in an evidence locker or for known reference material from herbarium specimens.

1. Dried plant matter (0.5 g) should be rinsed in 70% ethanol for 1 min to remove external bacteria and fungi, then rinsed three times in sterile water washes prior to air-drying on filter paper.
2. Dice plant matter into small segments (a few mm^2) and grind to an ultrafine powder in liquid nitrogen with a mortar and pestle.
3. Transfer powder to a centrifuge bottle with 5 mL extraction buffer (8M urea, 250 mM Tris-HCl [pH 8], 400 mM NaCl, 50 mM EDTA [pH 8], 0.5% 2-mercaptoethanol), and sterilize by filtration through a 0.45-μM filter. Mix by inversion.
4. Add solid SDS (1% wt/vol) and incubate at room temp for 60 min with occasional mixing.
5. Add 20% PVP (MW 10,000, –20°C) to a final concentration of 6%. Add ½ vol of 7.5 M ammonium acetate.
6. Incubate solution on ice for 30 min, centrifuge at 4000 rpm (Beckman Instruments model JA-20.1 rotor) for 15 min. Centrifuge further at 12,000 rpm for 10 min.
7. Transfer supernatant to sterile centrifuge bottles, then mix well with an equal volume of isopropanol by inversion and at room temp for 10 min.
8. Centrifuge solution at 12,000 rpm at 15°C for 20 min. Wash the pellet with 85% ethanol and dry in heat block set at 60°C for 20 min. Dissolve the pellet in 1.2 mL sterile water.
9. Centrifuge DNA in solution at 14,000 rpm in a bench-top centrifuge for 1 min. Transfer supernatant to a 1.5-mL microcentrifuge tube.
10. Perform a phenol:chloroform (1:1) extraction and then a chloroform-only extraction. Pellet nucleic acids by the addition of an equal volume of isopropanol at room temperature.
11. Rinse the pellet twice with 85% ethanol, dry in 60°C oven for 20 min, and resuspend in 100 μL of TE buffer (10 mM Tris-HCl [pH 8], 1 mM EDTA).

Some considerations when performing a DNA extraction are

- To prevent nuclease enzymes from digesting the DNA during the extraction procedure, samples frozen in liquid nitrogen should not be allowed to thaw. Proteinase K is a helpful additive to extraction buffers to digest proteins.

- For improved recovery of high molecular weight (high quality) DNA, samples should be handled gently to avoid shearing of the DNA strands. Vigorous pipetting and vortexing will damage the DNA and decrease the quality of the DNA recovered.
- Plant tissue samples ideally should be quick frozen in liquid nitrogen and stored at –70°C in a noncycling freezer. DNA may be susceptible to damage by exposure to prolonged heat, humidity, and ultraviolet radiation.

4.8 The Future of Plant Biology

Once model plant organisms have had their entire genome sequenced, the genes or open reading frames (ORFs) need to be assessed for protein function and for interactions with other proteins within the cell. With the imminent completion of the *Arabidopsis* genome-sequencing project, the following goals have been identified:

- Develop simple gene expression systems to study function of proteins (functional genomics).
- Create comprehensive sets of gene mutants to assess consequences of loss of function.
- Quantitate *in vivo* gene expression using microarray chip technology.
- Produce antibodies to all deduced proteins.
- Describe global protein profiles under various environmental conditions.
- Develop a global understanding of posttranslational modification of proteins.
- Create plant artificial chromosomes for ease of transfer of economically important traits.
- Identify regulatory sequences for all genes.
- Describe the three-dimensional structure of every plant-specific protein family.
- Develop comprehensive systems to better understand uptake, transport, and storage of ions and metabolites.
- Define a predictive basis for the conservation and diversification of gene function.
- Compare gene sequences between species and assess in an evolutionary context.
- Develop bioinformatic systems to visualize and predict better models for plant function.

In addition, many commercial biotechnology ventures are offering plant DNA-based services. These services include the use of capillary electrophoresis systems, pyrosequencing, microarray scanning, and bioinformatics databases to offer a myriad of services such as the following:

- High-throughput expressed sequence tags (EST) genotyping
- Single nucleotide polymorphism (SNP) discovery and characterization
- Microarray analysis
- DNA Bank support services
- Seed identification
- Plant variety identification
- Support for the policing of plant breeders rights
- Genetic purity analysis of plant products
- Food genetic purity analysis
- Forensic analysis of plants
- Plant breeding support

4.9 Summary

Plant reproduction, either through natural means, traditional breeding methods, or the use of new DNA technology, results in the creation of new generations of plants. Understanding the reproductive process is critical to the correct interpretation of DNA typing profiles and performing accurate estimations of random match frequencies within plant populations. Careful consideration and thorough research into the genetic history of a plant can save on confusion and time in determining the optimal manner for creating a comparative population database for botanical evidence that will be introduced into court for criminal and civil casework. Some key points to consider when considering plant evidence after species identification has been made and further DNA-based individualization methods are going to be performed include the following:

- Is the plant sexually or asexually reproduced?
- At the population level, is there more than one form of reproduction that could be occurring that might have relevance to the evidence at hand (e.g., cloning and seed reproduction combined)?
- Is there a significant level of inbreeding for this plant species? (This is important because inbreeding will result in higher percentages of shared DNA between plants than if they were randomly mating, obligate outcrossing plants).
- Above all else, do the DNA results indicate a possible match, common origin for clonal plants, exclusion, or are they inconclusive? Care in using the appropriate terminology for plant identification and individualization test results will aid in the acceptance of botanical evidence into the judicial system.

General References

Arntzen, C.J. et al., GM crops: science, politics and communication, *Nat. Rev. Genet.*, 4, 839, 2003.

Baucher, M. et al., Lignin: genetic engineering and impact on pulping, *Crit. Rev. Biochem. Mol. Biol.*, 38, 305, 2003.

European roundup, *Genetic Engineering News*, 20, 42, 2000.

Giovannetti, M., The ecological risks of transgenic plants, *Riv. Biol.*, 96, 207, 2003.

Giri, C.C. and Vijaya Laxmi, G., Production of transgenic rice with agronomically useful genes: an assessment, *Biotechnol. Adv.*, 18, 653, 2000.

Pasternak, J.J. et al., Development of DNA-mediated transformation systems for plants, *Biotechnol. Adv.*, 1, 1, 1983.

Pauls, K.P., Plant biotechnology for crop improvement, *Biotechnol. Adv.*, 13, 673, 1995.

Rong-Zhao, F. et al., Extraction of genomic DNA suitable for PCR analysis from dried plant rhizomes/roots, *BioTechniques*, 25, 796, 1985.

Safety and oversight of transgenic plants, *Genetic Engineering News*, 20, 24, 2000.

Shargool, P.D., Biotechnological applications of plant cells in culture, *Biotechnol. Adv.*, 3, 29, 1985.

Somerville, C. and Dangl, J., Plant biology in 2010, *Science*, 290, 2077, 2000.

DNA Structure and Function

Part 1: Fundamental Genetics

NICHOLAS CS YANG

Contents

5.1 Introduction

Humans begin as a single cell when the sperm fertilizes the egg, and inevitably grow to be a collective of 200 trillion cells, all functioning together to form organs and tissues within the human body. In reaching their present form in the body, many billions of successful cell divisions occurred by an equal division of genetic material and cytoplasm of the cell. Within the cell, a myriad of molecules are present, including DNA (deoxyribonucleic acid), RNA (ribonucleic acid), proteins, lipids, and carbohydrates. Genetics is a complex area of study focused on the inheritance of cell contents, and in particular DNA. In this chapter, the foundation for understanding genetics

0-8493-1529-8/05/$0.00+$1.50

will be presented in its simplest terms, but for a greater amount of detail refer to any standard biology or genetics textbook.

5.2 Deoxyribonucleic Acid (DNA) Structure

DNA (Deoxyribonucleic acid) provides the fundamental blueprint of genetic information for the cell. Discovered by Watson and Crick, the molecular model for a DNA molecule is a twisted double helix of two connected antiparallel strands that vary in length depending on the organism. The building blocks of DNA consist of four nucleotide bases — adenine, cytosine, guanine, and thymine — that are joined together by chemical bonds to form a molecule often depicted as a ladder. Attached to each nucleotide base are a phosphate group and a 5-carbon (pentose) sugar. Complementary base pairing (to form the steps of the ladder) occurs between each of the two DNA strands in a predictable fashion. For base-pairing, adenine only hydrogen bonds with thymine (in RNA, Uracil) and cytosine only hydrogen bonds with guanine.

Organisms store and use DNA by translating the sequence or order of nucleotide bases into the amino acids that compose critical proteins. Another nucleic acid is present in cells called RNA (Ribonucleic acid, usually single-stranded) and functions as an intermediary molecule during the translation process to read DNA-coded information for making proteins. DNA allows for storage of hereditary genetic information by using the four nucleotides in a particular sequence, and by following the A-T and C-G bonding rules. The order of the nucleotide bases in DNA is faithfully maintained during mitosis to give an exact copy of a cell after DNA replication and cell division. Any change in the order of the sequence may result in a mutation at the molecular level, the genotypic level, or the phenotypic level.

For example, a change at the molecular level can be interpreted as a nucleotide substitution that could affect the structure of a gene product (i.e., protein). For example, sickle cell anemia, a hereditary disease manifested by the presence of oxygen-deficient red blood cells, is caused by a single nucleotide substitution in the sequence of a gene. This one difference results in a change of structure from a donut to a crescent-shaped red blood cell, and limits the cell's oxygen transport function. A genotypic change refers to a heritable change in the DNA since a "genotype" is defined as the genetic complement or genome that is passed from parent to offspring, in which the function of the gene may not be affected. This nuance between molecular and genotypic change really amounts to whether or not a nucleotide difference is within the protein-coding region of a gene (an exon) or the DNA found within a gene that is noncoding (an intron). A phenotypic change is

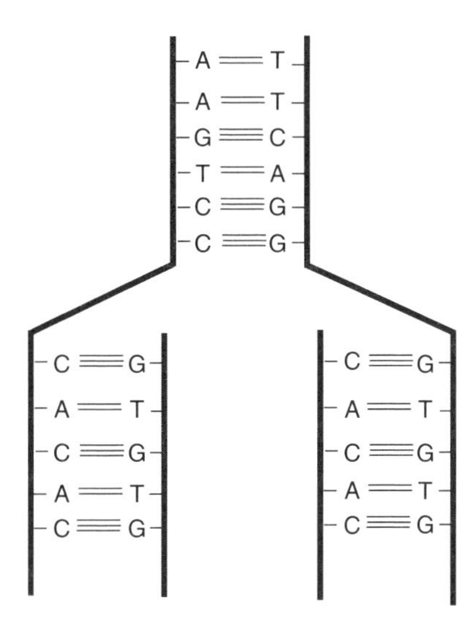

Figure 5.1 The semiconservative model of DNA replication has the two DNA strands of the parent molecule separating and each one functioning as a template for the synthesis of a new complementary DNA strand.

an alteration in an observable trait or characteristic such as the brown hair or blue eye color of an organism.

Structurally, the nucleotides are attached to each other in a 5' to 3' orientation to form the sides of the ladder (rather than the rungs) and are in opposite orientation; therefore, they are called antiparallel strands. These strands are also called "complementary" because of the base-pairing rules previously discussed and serve as templates for the replication of new strands of DNA. As a double-stranded DNA molecule separates into its single strands during replication, one of the original DNA strands will complementarily bind with a newly synthesized DNA strand, and the other original DNA strand will bind with the second newly synthesized strand. Thus, this process of replication is considered to be semiconservative (Figure 5.1). DNA does not exist as a long double-stranded chain in the cell as is so often depicted for ease of explanation in illustrations. In fact, DNA is packaged tightly with proteins and supercoiled (tightly wrapped like a ball of yarn) into a structure called chromatin in the nucleus of the cell. For further compression, the DNA is bound even more tightly into chromosomes in order to facilitate movement of large sections of the DNA during cell replication.

In most forensic laboratories where DNA testing is available, a common molecular biology technique called polymerase chain reaction (PCR) is used to copy segments of DNA. In many ways, the synthetic PCR process is similar to the natural DNA replication process in cells. The PCR reaction requires several components for a successful amplification of the target sequence:

- Taq polymerase enzyme
- Nucleotides
- Magnesium
- Buffer
- PCR primers
- DNA template

The PCR reaction is a three-step process that includes successive steps of denaturation, annealing, and elongation of the DNA. After all the ingredients have been added into a small volume tube and placed in a thermalcycler (a machine that rapidly cycles through temperature phases), DNA denaturation begins; the bonds between double strands break to produce single-stranded DNA. During the annealing step, the primer (a short sequence, ~15 to 40 nucleotides, of single-stranded DNA) binds to the complementary target sequence of the single-stranded DNA. The polymerase enzyme uses the primer as a foundation to catalyze the addition of free nucleotides (dNTPs) that are complementary to the sequence of the original single-stranded DNA template.

This process can be repeated as many times as necessary to generate the PCR product. After each cycle, the numbers of copies of the target DNA will double, resulting in an exponential growth rate for target sequences. After the first cycle, for example, one double-stranded DNA (dsDNA) molecule would become two dsDNA molecules. Following the second cycle, two dsDNA molecules would generate four dsDNA molecules. Theoretically, if one cell's DNA was extracted, and the PCR reaction was successful, then, 28 cycles would generate over 250 million molecules of the target sequence. Essentially, PCR can be equated to "molecular xeroxing" in which exact copies are made of the target DNA site, much like the replication paradigm within the cell. The difference between the synthetic PCR process and the natural process of cellular DNA replication lies in the amount of DNA that is being reproduced or "copied." In cellular replication, the entire cell content of DNA is duplicated, whereas in the PCR process, only targeted segments of DNA are reproduced.

5.3 Mitosis

Mitosis is a process that replicates the chromosome number; the chromosomes are later divided equally between the two identical daughter cells (Figure 5.2; see color insert following page 238). During the life cycle of a cell, mitosis consists of a short sequence of events, but each organism depends on this high-fidelity procedure to survive. During a stage called interphase, most cells spend their time growing and synthesizing some of the cellular machinery to undergo mitotic period division. The cell cycle consists of the three additional cell cycle stages designated as G1, S, and G2. The cell undergoes rapid growth during the G1 phase, which increases the cell size to almost twice the original volume. During the S phase, the genome is condensed into chromosomes, and the replication of DNA results in two exact copies of a cell's DNA. In the G2 phase, all of the final preparations are made for cell division, such as the appearance of the spindle fibers (composed of microtubules) and centrioles. Microtubules are structurally similar to the fibers of cilia and flagella in prokaryotes. The centrioles are structures located at each tip of the spindle apparatus and are similar in some respects to the basal bodies in prokaryotic flagella formation. Both microtubules and centrioles are necessary for chromosome movement during cellular replication.

The process of mitosis can be further divided into four distinct phases based on the physical appearance of the chromosomes: prophase, metaphase, anaphase, and telophase. In prophase, a profound change in genome organization is witnessed. The chromatin in the nucleus of the cell begins to shorten and coil to form a chromosome (a DNA structure consisting of two sister chromatids joined by a centromere). As the nuclear membrane begins to disappear, specialized microtubules called spindle fibers begin to form in the cytoplasm. In metaphase, the nuclear membrane disappears completely, and the chromosomes align themselves near the middle of the cell. These movements are based on the attachment of each chromosome through a specific point along its length, the centromere, to the spindle apparatus. In anaphase, the sister chromatids are pulled apart at the centromere by the spindle fibers and move to the opposite poles or sides of the cell. This phase is clearly visible as an act of "repulsion" as the chromosomes are dragged through the cell. In telophase, the chromatids begin to uncoil, the nuclei begin to reform, and a process called cytokinesis divides the cytoplasm. The result is two diploid daughter cells, each containing one exact copy of each of the chromosomes. At this point, the "daughter chromosomes" begin to relax and uncoil to revert back to the noncondensed interphase state. The nuclear membrane forms again and the nucleoli also form. In plant cells, a cell plate forms to separate the two cells; in animal cells, an indentation called the cell furrow is established.

5.4 Meiosis

Meiosis is similar to mitosis; however, each chromosome undergoes an extra cycle of division to produce four haploid cells instead of two diploid cells as in mitosis (Figure 5.3A and Figure 5.3B; see color insert following page 238). This process occurs in the specialized cells (testis/ovary) of sexually reproducing organisms during the process of fertilization. When fertilization occurs, two haploid cells are joined together to form a single diploid cell.

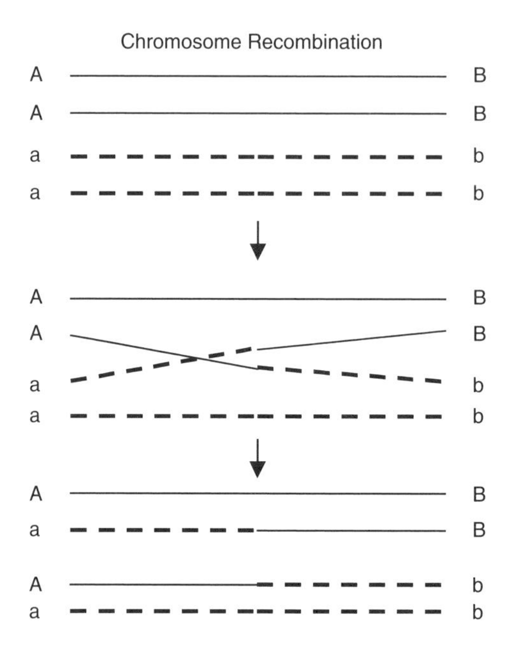

Figure 5.4 Sexual reproduction and the process of meiosis results in the incredible amount of genetic diversity present in a randomly mating plant population. Vegetatively propagated plants result in identical or highly similar genetic content between parent and offspring, and therefore have populations with low genetic diversity.

The process of meiosis and chromosome recombination results in genetic diversity, the billions of combinations that produce a unique genetic offspring (except for identical twins) each time sexual reproduction occurs (Figure 5.4). In meiosis, a 50% reduction in the number of chromosomes occurs. This is important because when an egg and sperm unite in the process of fertilization, they form a diploid zygote that will develop into an offspring that will have a unique combination of DNA derived from each parent. If the DNA content was not reduced by 50%, then each generation of offspring would experience a continuous doubling of DNA content, a condition called "polyploidy." A significant difference between animals and plants is that some plants can tolerate higher ploidy levels (increased DNA content) without an apparent loss of function while most animals cannot tolerate polyploidy.

There are two phases to the meiotic process, called meiosis I and meiosis II. Meiosis I reduces the number of chromosomes per cell and homologous chromosomes recombine to increase genetic variation. As the cell matures in interphase I, chromosomes are replicated. In prophase I, the chromosomes begin to coil and condense, the nuclear membrane also disappears, and spindle fibers slowly form. Homologous chromosomes begin to pair together so that genetic recombination (shuffling of DNA content between the chromosomes) can proceed. In metaphase I, the chromosome homologue pairs line up at the equator of the cell, and the centrioles and spindle fibers become more defined. In anaphase I, the homologues separate and migrate toward the opposite poles of the cell through the attachment of the spindle fibers to the center of the chromosome (the centromere). Without the centromere, homologous chromosomes may not be able to separate to opposite poles of the cell, and thus, a nondisjunction event would occur (referring to the uneven distribution of chromosomes to each daughter cell). In the final telophase I phase of meiosis I, the original cell divides into two daughter cells, with an equal separation of the cell contents (cytoplasm, organelles, etc.).

Meiosis II essentially begins at prophase II even though there is some uncoiling of the cell's DNA after telophase I. Chromosomes coil and condense while the centrioles travel to the poles and spindle fibers form. In metaphase II, the chromosomes migrate and line up at the equator of the cell. In anaphase II, the sister chromatids separate and move towards opposite poles within the cell while in telophase II, the chromosomes begin to uncoil, a nuclear membrane begins to form, and the cytoplasm is divided. After meiosis II is complete, the result is four haploid daughter cells, each containing one copy of each chromosome.

5.5 Mendelian Genetics

Dubbed the father of genetics, Johann Gregor Mendel uncovered the cellular mechanisms of inheritance through a carefully chosen set of experiments. He selected pea plants because they could self-fertilize (which means offspring remain fertile), they offer a number of easily scorable characteristics, and they have a relatively short generation period. He collected data from his experiments for over eight years to avoid statistical errors from small sample sizes and published in 1865. Mendel named the parental generation P1 and named the offspring the first filial generation designated as F1. The next generation was named F2 from the self-fertilization process of the F1 generation. If the parents were of two different pure-breed varieties (i.e., with a smooth or wrinkled seed appearance), the F1 generation presents only one phenotype of the parental trait (smooth). If the F1 generation were self-fertilized, the F2 offspring presents both traits from the parents but in an expected 3 (smooth):1 (wrinkled) ratio (Figure 5.5). From these experiments, Mendel reached the following conclusions:

- Plant characteristics or traits could be hidden or not expressed. Thus, he reasoned that traits must be either dominant or recessive. A dominant gene is one that can be expressed in the presence of another gene. So, P1 and F1 plants have the same phenotypes (observed physical properties) but different genotypes (genetic constitution).

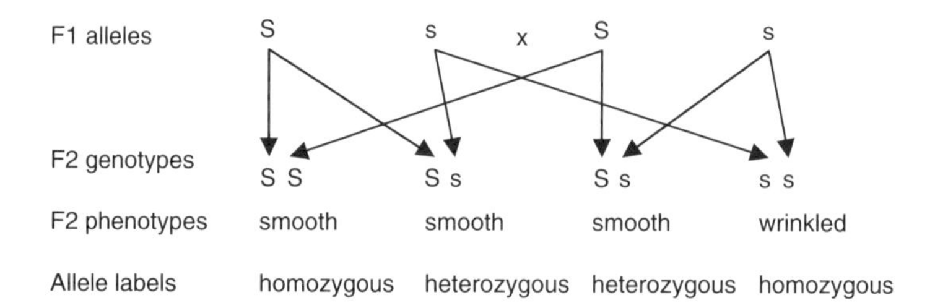

Figure 5.5 This figure schematically illustrates how classical genetics can be useful for predicting inheritance patterns for single dominant traits. A dominantly expressed and inherited trait is expected to be inherited in a 3:1 ratio when two heterozygous parents are crossed to yield offspring.

- He deduced that each generation carries two genes, or alleles, that determine seed shape. From these simple monohybrid crosses, Mendel reasoned that gene pairs must segregate from each other during gametogenesis; this is known as Mendel's First Law, the principle of segregation.
- Next, data from dihybrid crosses of smooth yellow seeds and wrinkled green seeds were collected. The F1 generation seeds were all smooth and yellow (dominant). The F2 generation seeds produced four different phenotypes of seeds: 9/16 smooth yellow, 3/16 wrinkled yellow, 3/16 smooth green, and 1/16 wrinkled green. These results illustrated that two different genes from the same parent can segregate independently of each other, and consequently, gametes of all combinations can be produced. This is known as Mendel's Second Law, the principle of independent assortment. The frequency of recombination is dependent on the physical distance between genes on the chromosome. However, if genes are located close together on a chromosome (50 Centimorgans or less), they will be inherited together (genetic linkage).

5.6 Population Genetics

Humans as well as plants can be distributed over large geographic areas, and can be clustered in local regions also known as populations. These populations share a common gene pool, or a set of alleles (genetic types). Because populations are dynamic (environmentally influenced), the frequency of an allele that is present within a population can change over time. Scientists can study the genetic structure of a population by examining allele frequencies as well as the stability of alleles over several generations.

Since Mendel's initial experiments with plants, many studies have examined the inheritance patterns of genes and their corresponding traits in humans. Godfrey Hardy and Wilhelm Weinberg independently arrived at the conclusion that Mendelian genetics does not differentiate between the modes of inheritance of a gene and the frequency of the dominant/recessive alleles in the population. Hardy and Weinberg developed a mathematical model to study genetics in populations using several assumptions:

- First, the population must be large enough that no errors in sampling for the allele frequencies will occur.
- Second, mating between individuals within a given population must be random.
- Third, no particular advantage must exist for any given genotype, thus, all genotypes have an equal chance to survive and be inherited.
- Fourth, the genetic mutations and migration patterns within the population that may change the frequency of any given allele are rare events and can be disregarded.

How does one calculate the allelic and genotypic frequencies using the Hardy–Weinberg model? While using a two-allele system, assume that the frequency of the dominant allele, S, is represented by p, and the frequency of the recessive allele, s, is represented by q. The sum of the alleles, $p + q$, should correspond to 100% of the alleles present in that gene pool ($p + q = 1$). Thus, the distribution of genotypes in the next generation is expressed as $p^2 + 2pq + q^2 = 1$, or $S^2 + 2Ss + s^2 = 1$.

The frequency of the dominant and recessive alleles in the gametes from the parental generation, P1, determines the frequency of the alleles and genotypes of the F1 generation. If the allele frequency of a gene in a population remains steady through many generations, it is said to be in a state of genetic equilibrium.

For example, assume that brown eyes are a dominant trait (BB) and green eyes are a recessive trait (bb) with 1 in 100 individuals having this distinctive phenotype. Using the Hardy–Weinberg model ($p^2 + 2pq + q^2 = 1$), the frequency of the "bb" genotype is q^2 (1 in 100 = 0.01 = q^2), and the frequency of q (or "b" allele) would be the square root of q^2, or 1 in 10 (10%). Ten percent of the population has the "b" allele, and that would mean $p = 1 - q$, or 90% of the population has the "B" allele. Based on these allele frequencies, we calculate the heterozygote allele frequencies in this population, 2pq or $2*0.9*0.1 = 0.18$ (18%). These results would indicate that 81% of people carry the "BB" gene, 18% of the population carry the "Bb" gene, and 1% carry the "bb" gene. Because brown eyes (BB) are a dominant trait, "Bb" is expressed as a brown eye phenotype, and not a combination of brown/green eyes. In addition, individuals with the "Bb" genotype are carriers for the "b" allele. Subsequently, 99% ($0.81 + 0.18 = 0.99$) of the population has brown eyes, and only 1% of individuals has green eyes. (See Table 5.1.)

Table 5.1 An Example of Allele, Genotype, and Phenotype Frequencies for Brown and Green Eye Color

Allele(s)	Allele Frequency	Genotype Frequency	Phenotype Frequency
BB	0.9	0.81	0.99
Bb	0.1	0.18	N/A
bb	N/A	0.01	0.01

With a parental cross-hybridization (P1) = **SS** x **ss.**
Let S = smooth and s = wrinkled:

	S	S
s	Ss	Ss
s	Ss	Ss

The first filial generation (F1) = Ss = all are smooth in phenotype.
For an F1 cross hybridization = **Ss** x **Ss:**

	S	s
S	SS	Ss
s	Ss	ss

The F2 generation demonstrates a 3:1 ratio of smooth:wrinkled phenotype.

$$p^2 + 2pq + q^2 = 1 \rightarrow (BB)^2 + 2(Bb) + (bb)^2 = 1 \rightarrow$$

$$(0.9)^2 + 2(0.9)(0.1) + (0.1)^2 = 1 \rightarrow 0.81 + 0.18 + 0.01 = 1$$

5.7 Conclusion

Part 1 of this chapter presented a brief and concise overview of genetics, which is required for a basic understanding of the important features of chromosome replication and cell division, including genetic processes common to both animal and plant eukaryotic cells. In some instances, exceptions to the genetic rules may exist depending on the organism under study, which may not fit exactly into a particular genetic archetype. Inherent in this chapter is the information that is critical for performing DNA analyses on samples in forensic science. Since many plants are under less genetic constraints than animal cells, as illustrated by the ability to tolerate polyploidy without a loss of function, a thorough understanding of the genetics for each plant system is necessary before reaching a logical conclusion.

For plant forensic applications, it is important for the scientist to understand the form of reproduction that the plant has undergone (e.g., cloning, grafting, seed propagation, inbreeding). This allows for the use of appropriate population genetic statistics to estimate the random match frequency of a DNA profile within a given population of a plant species. Also described in this chapter was a brief introduction to the study of population genetics and the importance of a population being in Hardy–Weinberg equilibrium in order to use the product rule (multiplication of allele frequencies from independently inherited genetic loci). In evidentiary samples that generate human DNA profiles from commercially sold kits (PCR-based technology), the product rule is also used to calculate the random match frequencies that give statistical significance to the sample.

Part 2: Molecular Biology Basics

HEATHER MILLER COYLE

Contents

5.8 DNA Extraction

There are many different methods for the extraction of DNA (including nuclear, mitochondrial, and chloroplast forms) and purification away from other cellular components. These methods include the use of organic solvent extractions, resin binding for purification (e.g., Chelex), and differential binding through silica support matrices. Each method has positive and negative aspects to the procedures employed for DNA isolation. Organic solvent extractions for DNA isolation are more time-consuming, and small amounts of sample are lost each time a solvent is added and the supernatant is removed. However, they are most optimal for the removal of a wide variety of potential PCR reaction inhibitors that might be included in forensic samples. Resin-binding methods for DNA purification are typically simple to perform but may not bind all forms of DNA equally well, especially if some of the DNA samples are degraded. Differential binding of DNA on an inert silica matrix followed by subsequent release of the DNA after several washing steps seems to yield high-quality DNA for PCR amplification without significant inhibitor effects. After any of these types of DNA extraction, the end result will consist of nuclear, chloroplast, and mitochondrial DNA in an aqueous solution. If further separation of the DNA forms is desired, differential centrifugation in a cesium chloride gradient may be sufficient.

5.9 PCR Amplification

The purpose of the PCR method is to make a large number of copies of a segment of DNA being tested. Amplification of additional copies of DNA is often necessary in order to have a sufficient starting template for DNA sequencing or fragment-length analysis. There are three major steps in a PCR reaction, and each of these three steps is repeated for 30 or 40 cycles. This repetitive process is performed with an automated cycler, which can heat and cool the tubes containing the reaction mixture in a very short time. The first step is called denaturation and occurs at 94°C. During denaturation, the double strands of a DNA molecule melt open to yield two single strands of DNA that are accessible to the PCR primers and enzymes.

The second step of the PCR process, annealing, occurs at approximately 54°C. Complementary base pairing between the single-stranded primers and the single-stranded template occurs as the short PCR primer sequences bind to their complementary section of the target DNA. The DNA polymerase (the enzyme used for building new strands of DNA) will attach to the short double-stranded section of primer-template and will begin to replicate the template to make an exact copy.

The final step of the PCR reaction is the extension step that occurs at 72°C. This is the optimal working temperature for the DNA polymerase enzyme. Free nucleotides are added in order by the DNA polymerase to create the duplicated strand. The nucleotide bases (complementary to the template) are coupled to the primer on the 3' side and the polymerase adds them in a 5' to 3' direction, reading the template from the 3' to 5' side, as bases are added complementary to the template.

Before the PCR product is used in further applications, it has to be checked to ascertain if a product has been formed. Not every PCR reaction may be successful for a variety of reasons. There is a possibility that the quality of the DNA is poor, insufficient DNA is present, that one of the PCR primers does not bind well to the template, or that there is an excess of starting template. All of these circumstances may result in an inefficient reaction that yields little or no PCR product. Typically the presence or absence of a PCR product is determined by loading a sample of DNA on an agarose gel, performing electrophoresis, and staining with ethidium bromide to visualize a product under ultraviolet light.

5.10 Detection of Genetic Polymorphisms

The ability to distinguish between individuals or different nonhuman samples is ultimately a result of the ability to measure genetic differences. These genetic differences may be represented as fragment-length differences due to the presence or absence of restriction sites (useful for restriction fragment-length polymorphisms [RFLPs] and amplified fragment-length polymorphisms [AFLPs]) or nucleotide sequence differences (used in DNA sequencing, polymarker [PM/DQA1] tests, or single nucleotide polymorphisms [SNPs]). What creates a fragment-length difference? Restriction enzymes are DNA-cutting enzymes found naturally in bacteria and are harvested from them for use in molecular biology. Because these enzymes cut within the DNA molecule, they are often called restriction endonucleases. A restriction enzyme recognizes and cuts DNA only at a particular sequence of nucleotides. For example, the bacterium *Hemophilus aegypticus* produces an enzyme named HaeIII that cuts DNA wherever it encounters the sequence

5'-GGCC-3'

3'-CCGG-5'

The cut is made between the adjacent G and C nucleotides. Treatment of DNA with a restriction enzyme produces DNA fragments, each with a precise length and nucleotide sequence. These fragments can be separated from one another on an agarose gel and the DNA sequence of each fragment determined if necessary.

Fragment-length differences can also be generated through differential amplification by PCR and those fragments labeled with a colorimetric or fluorescent dye "tag." Short tandem repeats (STRs) are repeated sequences of a few (usually four) nucleotides, for example, TCAT,TCAT,TCAT,TCAT, and are the most common form of fragment-length analysis in forensics today. The exact number of repeats (6, 7, 8, 9, etc.) varies in different people within the population and thus gives this form of DNA testing a high degree of discrimination power. Using ten different STR markers located on different chromosomes gives a random match probability or the chance that two people picked at random have the same DNA profile (except identical twins) as fewer than 1 in 3 trillion.

5.11 Tools of the Trade

5.11.1 DNA Cloning

Many people are familiar with the term "cloning" through the media coverage of reproductive cloning. There are different types of cloning, however, and cloning technologies can be used for other reasons besides producing a genetic twin of an organism. The three types of cloning technologies are

- Molecular cloning
- Reproductive cloning
- Therapeutic cloning

The transfer of a DNA fragment of interest from one organism to a self-replicating genetic element such as a bacterial plasmid for further study is often called molecular cloning. Plasmids are self-replicating, extra-chromosomal circular DNA molecules, distinct from the normal bacterial genome. The DNA of interest can then be propagated in a foreign host cell, typically through a rapidly replicating bacterium. This technology has been utilized since the 1970s, and it has become a common practice in molecular biology laboratories today. To "clone a gene," a DNA fragment containing the gene of interest is isolated from other chromosomal DNA using restriction enzymes and then united with a plasmid that has been cut with the same restriction enzymes. When the fragment of chromosomal DNA is joined with its cloning vector in the laboratory, it is called a "recombinant DNA molecule." Plasmids can tolerate up to 20,000 base pairs of foreign DNA. Besides bacterial plasmids, some other cloning vectors include viruses, bacteria artificial chromosomes (BACs), and yeast artificial chromosomes (YACs). Although bacterial plasmids are the most common cloning vectors, these other vector types are utilized when larger segments of DNA need to be replicated and studied.

Dolly, the sheep, was the first mammal to be created by a reproductive cloning process. By somatic cell nuclear transfer (SCNT), genetic material from the nucleus of a donor adult cell is transferred to an egg whose nucleus has been removed. Electric current or chemicals are required for cell division to occur and function to stimulate the reconstructed egg containing only nuclear DNA from the donor cell. Once the cloned embryo reaches a suitable stage of development, it is transferred to the uterus of a female host where it continues to develop until birth. Any animal created using SCNT is not actually an identical clone of the donor animal. While the donor and recipient nuclear DNA is identical, the mitochondrial DNA in the cytoplasm of the enucleated egg will be that of the recipient. Mitochondria, which are organelles within the cell, contain their own form of DNA. Acquired mutations in mitochondrial DNA are believed to play an important role in the aging process. Although hundreds of cloned animals are alive today, the number of different species is limited. Cloning attempts for certain species (e.g., monkeys, chickens, horses, and dogs) have been unsuccessful. It is thought that some species may be more resistant to SCNT than others. Substantial improvements in cloning technologies may be needed before many species can be cloned successfully.

Therapeutic cloning involves the production of human embryos for use in research studies. The goal of this process is to harvest stem cells that can be used for developmental studies and for novel disease therapies. Stem cells are important to biomedical researchers because they can be used to regenerate any type of specialized cell in the human body. Many researchers anticipate that future uses of stem cells will treat heart disease, Alzheimer's disease, cancer, and many other diseases. In the future, therapeutic cloning may be used to generate tissues and organs for transplantation. The stem cells would be used to generate an organ or tissue that is a genetic match to the recipient. This would reduce the risk of graft and organ rejection. However, many technical difficulties must be overcome before cloned organ transplantation becomes a routine medical procedure.

5.11.2 DNA Sequencing

DNA sequencing involves the deciphering of the order of nucleotide bases in a sample of DNA. The most popular method for doing this is called the dideoxy method, where DNA is synthesized from four deoxynucleotide triphosphates. The dideoxy method gets its name from the critical role played by synthetic nucleotides that lack the hydroxyl group at the 3 end of a carbon atom. A dideoxynucleotide (e.g., dideoxythymidine triphosphate, or ddTTP) can be added to the extending DNA strand, but when addition occurs, chain elongation stops because there is no 3 hydroxyl group for the next nucleotide to be linked to. Due to a stop in the elongation process, the dideoxy method is also called the chain termination method.

The DNA to be sequenced is prepared as single-stranded DNA and added to a mixture of the following:

- All four normal (deoxy) nucleotides in ample quantities (dATP, dGTP, dCTP, dTTP)
- A mixture of all four dideoxynucleotides, each present in limiting quantities and each labeled with a "tag" that fluoresces a different color (ddATP, ddGTP, ddCTP, ddTTP)
- DNA polymerase I enzyme
- Buffer solution with salts

Because all four normal nucleotides are present, chain elongation proceeds normally until, by chance, DNA polymerase inserts a dideoxy nucleotide instead of the normal deoxynucleotide. With a high ratio of normal nucleotides to dideoxy nucleotides, some DNA strands will extend for a few hundred nucleotides before the random insertion of a dideoxy nucleotide results in chain termination. At the end of an incubation period, the different-sized fragments are separated by length using the process of electrophoresis. The resolution of gel electrophoresis is so good that a difference of one nucleotide is sufficient to separate one DNA strand from the next shorter and next longer strand, allowing for the exact order of nucleotides to be determined. Each of the dideoxynucleotides is tagged with a different fluorescent dye when illuminated by a laser beam and is recorded by computer software to yield a "DNA sequence."

5.12 A History of DNA — Timeline

1869 — DNA is first isolated (Miescher).

1927 — Genetic damage is proven to be heritable when Muller showed that x-rays cause mutations that are passed from one generation to the next.

1928 — Transformation, the process of uptake of genetic material by a living organism, is discovered. Griffith injects mice with a mixture of live, avirulent, rough *Streptococcus pneumoniae* Type I and heat-killed, virulent, smooth *S. pneumoniae* Type II and observes that this mixture leads to the death of the mice. Live, virulent, smooth *S. pneumoniae* Type II bacteria are recovered from the dead mice, which suggests that genetic information from the heat-killed virulent strain has somehow been transferred to the avirulent live strain.

1941 — The "one-gene–one enzyme" hypothesis is proposed. Beadle and Tatum publish experimental results with the fungus *Neurospora crassa.* They conclude that a mutation in a gene controls the synthesis of an enzyme critical for essential nutrition in the fungus and the defective gene is inherited in a typical Mendelian fashion. These results ultimately lead to the "one gene–one enzyme" concept of biology, and Beadle and Tatum, along with Lederberg, are awarded the Nobel Prize in Medicine and Physiology in 1958.

1943 — Luria and Delbruck demonstrate that mutations occur randomly and that inheritance in bacteria is subject to natural selection. Delbruck, Luria, and Hershey are awarded the Nobel Prize in Medicine and Physiology in 1969.

1945 — Luria and Hershey demonstrate that bacterial viruses (bacteriophages) can mutate, which suggests the possibility that other virus types (such as the causative agents of influenza and colds) can also mutate.

1946 — Lederberg and Tatum demonstrate the sexual exchange of genetic material in bacteria.

1950 — McKlintock publishes proof that mobile genetic elements (transposons) exist in corn. She wins the Nobel Prize in 1983 for this work.

1952 — Lederberg and Zinder describe transduction, which is the transfer of genetic information by viruses. Dulbecco shows that an animal virus can produce areas of cellular lysis called plaques.

1953 — Crick and Wilkins, along with Watson, propose the double-helix structure of DNA. The proposed structure is based on x-ray crystallography studies on DNA performed by Franklin. Crick, Wilkins, and Watson are awarded the Nobel Prize in Medicine and Physiology in 1962.

1957 — Benzer shows that recombination can occur between mutations in the same gene and that genes are comprised of linear arrays of subunits that can be altered. Jacob and Wollman present evidence of the circular nature of the *Escherichia coli* chromosome by analyzing the results of interrupted mating experiments between conjugating bacteria. Kornberg discovers DNA polymerase.

1958 — Messelson and Stahl use density gradient centrifugation to demonstrate that the two strands of DNA unwind during replication and combine with newly synthesized daughter strands as predicted by Watson and Crick.

1959 — Sawada shows that antibiotic resistance can be transferred between *Shigella* and *E. coli* strains by the exchange of extrachromosomal plasmids.

1960 — Kornberg demonstrates new DNA synthesis in the absence of intact cells and later shows that DNA polymerase is the specific enzyme necessary to perform the synthesis.

1961 — Hall and Speigleman show that DNA isolated from T2 bacteriophage DNA can form hybrids with RNA, which leads to the later development of DNA hybridization methods commonly used in molecular biology. At this time, Crick, Brenner, and others propose the existence of a transfer RNA molecule that utilizes a three base code and acts in the direct synthesis of proteins. Nirenberg and Matthaei begin a series of ultimately successful experiments to decipher the genetic code. Nirenberg, along with Holley and Khorana, is awarded the Nobel Prize in Medicine and Physiology in 1968.

1967 — Szybalski and Summers show that only one strand of DNA (the sense strand) acts as a template for the transcription of RNA from a DNA template into protein.

1970 — Smith and Wilcox show that a restriction enzyme isolated from *Haemophilis influenzae* has the ability to recognize and cleave a specific DNA sequence. At this time, Temin and Baltimore independently report the discovery of reverse transcriptase in RNA viruses. Reverse transcriptase is an enzyme that uses single-stranded RNA as a template for the production of a single-stranded DNA complement to that RNA. This discovery demonstrates the possibility of a flow of genetic information from RNA to DNA. Temin, Baltimore, and Dulbecco are awarded the Nobel Prize in Medicine and Physiology in 1975.

1972 — Berg reports the construction of a recombinant DNA molecule comprised of viral and bacterial DNA sequences. Along with Gilbert and Sanger, Berg is awarded the Nobel Prize in Chemistry in 1980.

1973 — Cohen, Chang, Helling, and Boyer demonstrate that DNA digested with restriction endonucleases and combined with similarly digested plasmid DNA results in recombinant DNA molecules that can replicate in host bacterial cells.

1975 — Southern describes a new biological tool involving the transfer of restricted DNA fragments from a gel matrix to a nitrocellulose membrane for the detection of identifying DNA band patterns.

1977 — Chow, Roberts, and Sharp show that eukaryotic organisms have genes that are not continuous but are interspersed with stretches of nonprotein coding sequence. Roberts and Sharp are awarded the Nobel Prize in Medicine and Physiology in 1993. Also in 1977, Gilbert and Sanger develop independent methods for determining the exact nucleotide sequence of DNA. Along with Berg, Gilbert and Sanger are awarded the Nobel Prize in Chemistry in 1980.

1980 — Molecular biology kits become available as the commercial biotechnology industry begins to flourish.

1981 — Eli Lilly Corporation receives FDA approval to market the first recombinant protein, human insulin, for the treatment of diabetes.

1988 — Mullis introduces the polymerase chain reaction (PCR), a novel method for amplifying a specific DNA fragment from a very small amount of source DNA. PCR revolutionizes modern biology and has widespread applications in forensics, diagnostics, and gene expression analysis. Mullis is awarded the Nobel Prize in Chemistry in 1993.

1990 — The Human Genome Project formally begins an effort to identify all the genes on every chromosome in the human cell. The first approved gene therapy is performed with some success.

1992 — The entire 315,000-base nucleotide sequence of one of the 16 chromosomes of the yeast *S. cerevisiae* is published.

1999 — Researchers in the Human Genome Project reported the complete sequencing of the DNA making up the human chromosome designated chromosome 2.

5.13 Summary

The more modern tools of molecular biology are based on the traditional experiments that define human genetics and the study of heredity. This marriage of new and old technology is illustrated by the manner in which these topics have been combined in this two-part chapter. A complete understanding of the basic genetic principles is helpful when considering the reproductive strategies of plants that may be used in forensic applications. In addition, a command of the basic terminology used in molecular DNA analysis is useful when communicating the results of DNA testing between different agencies representing law enforcement, scientists, and legal counsel.

General References

Alberts, B. et al., *Molecular Biology of the Cell*, 3rd ed., Garland Publishing, New York, 1994.

Cummings, M., *Human Heredity: Principles and Issues*, 5th ed., Brooks/Cole Co., New York, 2000.

Lewin, B., *Genes V*, Oxford University Press, New York, 1994.

Lodish, H. et al., *Molecular Cell Biology*, 4th ed., W.H. Freeman and Co., New York, 2000.

Raven, P.H. and Johnson, G.B., *Biology*, 3rd ed., Mosby Year Book, St. Louis, 1992.

Plant Diversity

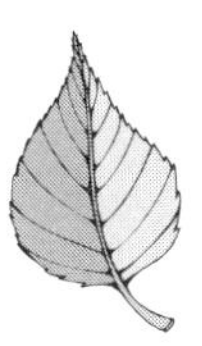

HEATHER MILLER COYLE

Contents

6.1 Introduction

Plants have both common and scientific names, with the latter generally based on a Latin derivation. For example, red oak is the English common name for *Quercus rubra.* A Swedish naturalist, Carl von Linnaeus, created a classification scheme for all living organisms consisting of a genus and species

0-8493-1529-8/05/$0.00+$1.50

Table 6.1 Ornamental Grasses

Name	Plant Type	Description
A. smithii (western wheatgrass)	Hardy perennial	2 feet high, foliage blue, seeds turn to beige
A. nebulosa (cloud grass)	Hardy annual	1.5 feet high, foliage green
A. donax (giant reed grass)	Hardy perennial	With flower plumes, can reach 14 feet in height
A. gerardii (bluestem)	Hardy perennial	7 feet, green foliage turns orange in the fall
A. sterilis (animated oats)	Hardy annual	3 feet tall, green foliage
B. minima (quaking grass)	Hardy annual	2 feet tall, green plants
C. morrowii (Japanese sedge)	Hardy perennial	1 foot high, green and white with curved leaf blades
C. lachryma-jobi (Job's tears)	Semi-hardy annual	2–4 feet high, green foliage
C. selloana (pampas grass)	Hardy perennial	Huge plumes reaching 14 feet tall
C. citratus (lemongrass)	Tender perennial	Aromatic, 6 feet, green foliage
C. papyrus (papyrus)	Tender perennial	Rhizomes, up to 8 feet high

name. In 1735, binomial (two-name) nomenclature was published and adopted as an international form of taxonomy. If similar, two organisms may share a genus but will have their own species name. The definition of a species that is generally accepted is the biological definition where two similar organisms mate and can form viable offspring. Plants, however, do not "fit" this definition of species as well as animals. Many exceptions exist because of the ability of plants to reproduce vegetatively. All organisms are classified by relationship into kingdom, phyla, class, order, family, genus, and species. For practical purposes, plants are generally known by a "common name," as well as a botanical Latin name consisting of genus and species. In addition to the Latin name, many cultivated plants have the same genus and species designation but an additional "cultivar" or "variety" name based on distinct morphological features. The concept of a cultivar or variety is analogous to the concept of human race as all humans are *Homo sapiens* but then can be further classified into Caucasian, African American, and Hispanic groups. Some common plants used in landscaping and gardening are listed with their binomial and Latin names (see Table 6.1 to Table 6.7).

6.2 The Plant Kingdom Classification System

6.2.1 Extant and Extinct Species

Certain types of information and inferences may be made when examining fossilized (extinct) and living (extant) plant species. Comparing today's living plant species with the fossil record is often difficult because so much information has not been preserved. Gaps in the fossil record exist, and we can

Table 6.2 Common Ground Cover Plants

Name	Soil Type	Description
Potentilla species (cinquefoil)	Rich, well-drained	2-foot-high bushy plants with strawberry-like foliage
Coronilla varia (crown vetch)	Poor to average soils, pH 6.5–7.0	2-foot-high, spreading green plants with pink flowers in June through August
Deutzia gracilis (deutzia)	Average to moist	2–4 feet deciduous shrubs, white flowers in May
Erica species (heather)	Moist, acid	1–2 feet, low shrubs, small leaves, colorful foliage possible (pink, purple, white)
Hedera helix (English ivy)	Any type	Evergreen, climbing, green and/or white foliage, mixed
Juniperus species (creeping juniper)	Fertile, rich, well-drained, normal to acidic	Low growing, prostrate, blue-green foliage, small needles
Convallaria majalis (lily of the valley)	Rich, loamy, moist	Oval, pointed leaves, 1 foot tall, white/pink flowers in spring

Table 6.3 Common Lawn Grasses

Name	Type	Exposure
A. affinis (carpet grass)	Creeping, warm weather	Full sun to part shade
B. dactyloides (Buffalo grass)	Creeping, warm weather	Full sun, requires dryness
E. ophiuroides (Centipede grass)	Creeping	Full sun
F. arundinacea (Tall fescue)	Tufting	Full sun to moderate shade
F. ovina (Hard fescue)	Tufting	Full sun
F. rubra (Fine fescue)	Creeping	Full sun to moderate shade
L. multiflorum (annual ryegrass)	Tufting	Full sun to partial shade
L. perenne (perennial ryegrass)	Tufting to semi-creeping	Full sun to partial shade
P. pratensis (bluegrass)	Creeping	Full to almost full sun
Zoysia species (Zoysia grass)	Creeping	Full to almost full sun

only surmise environmental conditions and causes of extinction based on the few pieces of the puzzle that have been preserved. Some of the earliest blue-green algae and bacteria were not well represented in the fossil record because they had soft bodies that did not lend themselves to fossilization. These types of early fossils date back to 3.5 billion years ago.

6.2.2 Creating Phylogenies

Identifying evolutionary relationships or lines of descent is called phylogenetic analysis. The most basic ancestral forms (primitive) are typically at the base of the phylogenetic tree, and intermediate and advanced forms progress as one constructs the tree upward. How can one classify an organism? The classic animal, mineral, and plant groups are sufficient to start. However, when one begins to encounter large numbers of similar plants then the sorting process must become more refined. Scientists typically consider external morphology

Table 6.4 Common Vines

Name	Propagation	Description
Celastrus scadens (American bittersweet)	Seed, softwood cuttings, root cuttings, suckers	Male and female plants, requires support and good soil
Bougainvillea	Stem cuttings	Climbing with dense flowers, fragrant
Aristolochia durior (Dutchman's pipe)	Seed, semi-hardwood cuttings	Flowers in summer
Dioscorea batatas (cinnamon vine)	Seed, cuttings, tuberous root pieces	Requires moist, well-drained soil, cinnamon scented flowers
Bignonia (crossvine)	Seed, softwood cuttings	Dense yellow-orange leaves
Humulus lupulus (common hops)	Seed, softwood cutting, root division	Fragrant flowers, dense climbing foliage
Ipomoea purpurea (morning glory, moon flower)	Seed, cuttings, root division	Fruit yellow-green, large flowers
Passiflora incarnata (passionflower)	Seed	Flowers open in morning, close in afternoon, very distinctive

Table 6.5 Common Shrubs

Name	Propagation	Flowers/Fruit
C. speciosa (common flowering quince)	Seed, cuttings	Flowers and fruit in spring
Forsythia species	Seed, cuttings	Flowers early spring
Ilex cornuta (Chinese holly)	Seed but more often terminal cuttings	Large red berries in fall
Prunus laurocerasus (English laurel)	Seed, cuttings	Flowers in spring
P. coronarius (sweet mock orange)	Seed, cuttings	Flowers early summer
Picea abies (Norway spruce)	Seed, cuttings	Cones
Taxus baccata (English yew)	Seed, hardwood cuttings	Red berries in fall
N. domestica (Nandina)	Seed, cuttings	Flowers, red berries in mid-summer

(see Figure 6.1) and internal anatomy, as well as DNA and protein relationships. Sometimes different classification strategies do not result in identical phylogenetic constructs (e.g., morphology versus protein comparisons).

Originally, organisms were sorted into two proposed kingdoms known as Plantae and Animalia. Occasionally, organisms were encountered that did not easily fit into one kingdom or the other (e.g., nonmotile sponges). With the invention of the light microscope, a new world of organisms became visible.

Table 6.6 Modern Garden Roses

Type	Growth Habit	Variety Names
Climbers	6–30 feet, upright or spreading	Blaze (red), Golden showers (yellow), Joseph's coat (orange-yellow)
English roses	Large bushes with mounds of flowers	Othello (deep pink), Bianca (white)
Floribunda	Stiff, twiggy, 2–4 feet tall	Cherish (pink), Showbiz (red), Iceberg (white)
Grandiflora	Large bushes, 4–8 feet tall	Viva (red), Queen Elizabeth (pink)
Hybrid Tea	Glossy but sparse leaves, 3–5 feet tall	Peace (creamy yellow), Mister Lincoln (red)
Miniatures	Similar to hybrid teas only in miniature, 12–24 inches tall	Cupcake (pink), Party Girl (light apricot), Winsome (lavender)
Tree roses	3 feet tall, grafts of hybrid tea or floribunda roses	Most varieties of hybrid tea or floribunda are used, grafts made at the nursery

Table 6.7 Common Fruits

Name	Propagation	Type
Blackberry (*Rubus*)	Tip cuttings	Perennial
Cranberry (*Vaccinia*)	Tip cuttings	Perennial
Currant (*Ribes*)	Stem cuttings	Perennial
Elderberry (*Sambucus*)	Seed or division of whole plants	Perennial
Fig (*Ficus*)	Stem cuttings	Perennial
Gooseberry (*Ribes*)	Stem cuttings	Perennial
Grape (*Vitis*)	Stem, leaf, or root cuttings	Perennial
Juneberry (*Amelanchier*)	Seed, stem cutting, or whole plant division	Perennial
Raspberry (*Rubus*)	From suckers or tip layers	Perennial
Strawberry (*Fragaria*)	Runners or rhizomes	Perennial
Watermelon (*Citrullus*)	Self-sown seed	Annual

To categorize single-celled organisms, a third kingdom, Protista, was created. To further accommodate some significant differences between prokaryotic and eukaryotic organisms, still more kingdoms were defined (e.g., Monera and Fungi). Monera include blue-green algae and bacteria. Endosymbiotic organisms, which are intimately interrelated out of life necessity, do not fit into any of these simple categories. The chloroplast, an organelle in plant cells responsible for photosynthesis, is thought to have an endosymbiotic origin. It has been theorized that the chloroplast began as a single-celled organism that was engulfed and took up permanent residence in a larger multicellular organism. In the 1960s, scientists began classification based on prokaryotes (single-cell organisms) and eukaryotes (multicell organisms). The prokaryotes have been the most difficult to classify as they are more ancient evolutionarily and are thought to, perhaps, be comprised of two distinct ancient groups, all of which come from some very ancient progenitor.

Figure 6.1 The distinctive shape, edges, vein patterns, and color of leaves can be useful for the classification of plant species.

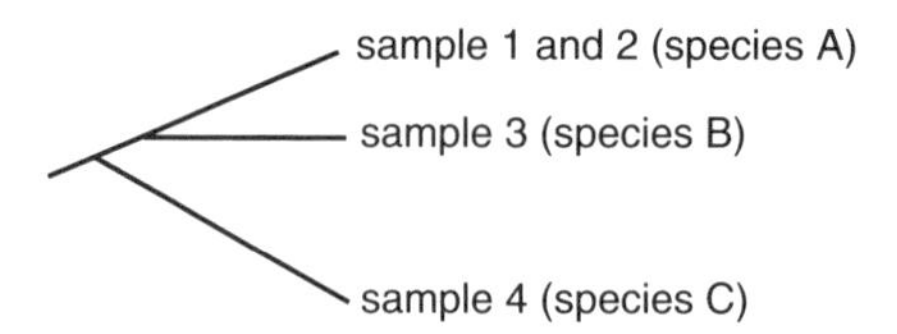

Figure 6.2 Phylogenetic trees are illustrations that depict evolutionary relationships between different plant species.

The pictorial arrangement of plant groups is generally referred to as a phylogenetic "tree" (see Figure 6.2). A taxonomic group is said to be "monophyletic" if a single ancestor gave rise to all of the species within that taxonomic group. A taxonomic group is "polyphyletic" if two or more ancestors gave rise to members of that group and the ancestral forms are not common to all members of the group. Assigning the appropriate taxonomic group to a species may be difficult due to homology and analogy of traits used for classification. Convergent evolution, for example, has resulted in birds and insects evolving similar or analogous equipment for flight although they do not have a common origin or ancestor from which this trait is derived. Homologous traits are those that were derived from a common ancestral member. As a general rule, if two species share a large number of homologous traits, it is likely that they share a common ancestor; however, exceptions to the rule do exist.

In plants, the term "division" is used to subdivide the major groups within the plant kingdom. The following is a commonly accepted classification scheme for plants:

- Nonvascular Plants
  Division Bryophyta (mosses) — Bryophytes are characterized by a waxy cuticle that helps the plant body retain water. Most are commonly called "mosses," and their habitat is typically moist and shady. Bryophytes are on average a few centimeters in height and typically grow horizontally as large mats across the surface of the ground.
  Division Hepatophyta (liverworts) — Liverworts are often grouped in with the mosses as they are quite similar in many respects; however, they may be even less conspicuous. They are classified by taxonomists into a separate group because their reproductive cycle is different than that of the Bryophytes.
  Division Anthocerophyta (hornworts) — Hornworts, again, are similar to the mosses but have some differences in the manner in which their sporophytes grow. In addition, their photosynthetic cells have one large chloroplast as opposed to the many smaller ones encountered in other types of plants.

- Vascular Plants (seedless plants)
  Division Psilophyta (whiskferns) — These plants are not the true ferns (Pterophyta) commonly encountered. One of the two genera in this division, *Psilotum* is common to the tropics and subtropics.
  Division Lycophyta (club mosses) — Many of these plant species grow as epiphytes (which means they use an organism as a living substrate but do not harm the organism). Club mosses are commonly found in woodland areas and the forest floors in the northeastern U.S.
  Division Sphenophyta (horsetails) — This is an ancient lineage of plants for which approximately 15 species of a single genus, *Equisetum*, remain. Horsetails are widely distributed but especially common to the damp regions of stream banks in the Northern Hemisphere.

Division Pterophyta (ferns) — This division is the most widely represented in our modern flora. Most ferns have leaves or fronds that grow coiled at the tip and unfurl as the leaf elongates. When these young fronds are harvested for gourmet salads, they are often called "fiddleheads." Ferns are generally restricted to moist, partially shaded habitats.

- Vascular Plants (seed plants, Gymnosperms)
  Division Coniferophyta (conifers) — Conifers are characterized by their reproductive structures called "cones." Within the conifer division are the pines, firs, larches, yews, spruce, cedars, junipers, cypresses, and redwoods. Most conifers are evergreens, which means they do not lose their leaves during the winter months. The majority of conifers are found in the Northern Hemisphere.
  Division Cycadophyta (cycads) — These plants resemble palms; however, they are not true palms, which are actually flowering plants.
  Division Ginkgophyta (ginkgo) — This division is represented by a single surviving species that is often called a maidenhair tree and is characterized by its fan-like leaves that are relinquished in the fall. The ginkgo is a common ornamental in urban cities since it can tolerate significant air pollution.
  Division Gnetophyta (gnetae) — This division consists of three genera: *Gnetum* (tropical vine), *Ephedra* (shrub in arid desert), and *Welwitschia* (a West African plant with extremely large strap-like leaves).

- Vascular Plants (seed plants, Angiosperms)
  Division Anthophyta (flowering plants) — Members of this division are extremely diverse and geographically widespread, and account for the majority of large temperate species. Approximately 235,000 species are known as compared to a little over 700 gymnosperm species (see Figure 6.3).

6.2.3 Plant Phylogenetic Resources

Plant phylogenic schemes are numerous, complex, and frequently confusing to the individual who may simply be interested in, for example, identifying the close relatives of a particular plant species. With this in mind, many resources have been developed to make plant taxonomy more convenient for the user. Of course, traditional phylogenetic and systematic groupings of plants are historically available in the botanical literature, but many resources are also now available over the Internet. Some of these resources include

Figure 6.3 Angiosperms, or flowering plants, dominate the plant kingdom.

- TreeBASE, a searchable database of phylogenetic trees maintained by Harvard University
- Tree of Life, an informative website that is attempting through funding by the National Science Foundation and the Green Plant Phylogenetic Coordination Group (GPPCG, also known as "Deep Green") to address fine scale phylogenetic classifications in order to answer some interesting evolutionary questions, such as providing an estimate for the number of times algae has colonized dry land
- Phylogenetic Resources, a listing of meetings, organizations, software, journals, etc. that is compiled by the University of California Berkeley Museum of Paleontology
- Deep Gene, a National Science Foundation project that is integrating plant phylogenetic information with plant genomics data
- The Green Tree of Life, another good resource for researchers and teachers for reconstructing the relationship of green plants with genomic information

6.3 Case History: Tracking the History of Grape Cultivars

Wine has been produced from grapes for over 5000 years, and a great deal of the history of plant breeding and wine production has not been carefully recorded or preserved. Some grape-breeding programs are carefully planned and maintained; for example, Cabernet Sauvignon, one of the world's most successful red wine cultivars, is the progeny of two classic Bordeaux varieties: Sauvignon blanc and Cabernet franc. At the University of California, Davis, scientists have deciphered the genetic heritage of 18 grape varieties grown in the northeastern region of France. Aided by a computer program, scientists Carole Meredith and John Bowers sorted the 18 grape varieties based on microsatellite patterns. The microsatellites were generated in much the same manner as human DNA profiles are generated, with short tandem repeat sequences (STRs), and an analogous paternity/maternity test was performed. Interestingly, 16 of the 18 varieties have been traced back to a single pair of prolific parents: Pinot and Gouais blanc. The Pinot variety is unquestionably a grape that yields a fine Burgundy wine. The Gouais blanc, however, was banned from use in France (multiple times) due to its inferior white wine quality. Grapevine selection is a time-consuming and expensive process, so if better breeding strategies can be developed based on clarified genetic relationships of cultivars through the use of microsatellite markers, the wine-making industry, too, will have benefited from new plant DNA technologies.

6.4 Case History: Tracing the History of the Potato

Potatoes have been in cultivation for an extended period of time. Art depicting potatoes and potato harvest gods dated as early as A.D. 400 has been recorded from the remains of the Nazca culture. Potatoes as we know them today can be traced to the Andean Mountains of South America where the Aymara Indians developed over two hundred varieties at elevations reaching 10,000 feet on the Titicaca Plateau. The Aymara found that freeze-dried potatoes were storable up to four years after "processing" using the following procedure:

- Spread potatoes on ground during frost season and cover with straw to protect from sun exposure.
- After several nights of freezing, compress by trampling to eliminate moisture and potato skins.
- Rinse for several weeks in running water to remove bitterness.
- Air-dry for 2 weeks and store for up to 4 years.

In 1553, Spanish Conquistador Pedro Cieza de Leon recorded the first written history of potatoes. Potatoes were soon brought to Europe, and slowly potato cultivation spread throughout the Old World. At first, potato cultivation was not embraced by many cultures because it was realized that the potato is a member of the nightshade family and the leaves and improperly grown tubers can contain alkaloids. In the 1600s, potato crop cultivation was encouraged by Prussian ruler Frederick the Great and a French agriculture specialist, Antoine Augustin Parmentier. Encouragement was provided through the use of threats of physical harm in the case of Frederick the Great; however, through his efforts potatoes maintained the people of Prussia through the Seven Years War (1756–1763). Parmentier achieved his goal via a marketing strategy to King Louis XVI and Marie Antoinette that spilled over into the French populace. European immigrants introduced the potato to North Americans throughout the 1600s; however, they did not become a popular food until the mid-1700s. Even when grown in large amounts, they were often used as animal feed rather than for human consumption.

In the mid-1800s, monoculture cultivation of potatoes in Ireland, and the resultant susceptibility of the crop to harvest failure, was the direct cause of mass population migration and untold suffering during the Great Famine of Ireland. At that time, Irish peasants were consuming large quantities of potatoes (approximately 80% of their diet). The climate in Ireland had proven to be ideal for potato cultivation for the past one hundred years, and the Irish people had become heavily reliant on this one crop. Three years of late blight, a fungal disease caused by *Phytophthora infestans*, had decimated the potato crops, and more than one million people starved to death. In addition, approximately two million Irish people emigrated and thus Ireland lost almost 25% percent of its entire population during this time period.

The potato became increasingly popular when Luther Burbank, an American horticulturist, developed through traditional plant breeding methods a potato cultivar called "Russet Burbank." From an "Early Rose" parental cross, 23 potato seedlings were generated. One of these seedlings produced potatoes of large size (2 to 3 times bigger) relative to all other potato cultivars of the time. In an additional test, a disease-resistant plant was identified that yielded a potato with a rough skin. By the late 1800s, Russet Burbank was marketed and grown all across the West, but Idaho became most famous for its potato cultivation. After a while, however, potato production began to decline and farmers realized it was due to their practice of replanting fields with culled potatoes (i.e., poor-quality potato "leftovers" not fit for market). A seed certification program was then established in Idaho, where potato crops were replanted from the best plants rather than from the culled discards not fit for market. Washington State is also recognized for superior potato cultivation, which is achieved through good irrigation, a long growing season, and rich volcanic soil. There, less than 500 farmers can produce upward of 18% of the U.S. potato crop. Although the potato suffered a brief dip in popularity in the 1950s because it was thought to be laden with fat, it still remains a staple of the American diet today. The Institute for Genomic Research (TIGR) is undertaking a large-scale genomic sequencing project for the potato.

6.5 Case History: The Origin of Rice

The origin of rice has been a point of debate for some time, primarily because crops that have been in cultivation for an extensive period of time are known to be valued and transported by humans throughout the world. This practice can make it difficult to pinpoint the first place rice was in cultivation and to identify the wild plant ancestors from which today's rice cultivars have originated. Some assumptions about water availability, fertile soil conditions, and ease of obtaining seed ("shattering" varieties) have been made and suggest that rice domestication may have first occurred in a basin area of Thailand, China, or Assam. Rice in cultivation is one of two different species, either *Oryza sativa* or *Oryza glaberrima. Oryza sativa* has been further subdivided into seven forms from different geographic locations:

- Two forms endemic to Africa but not in cultivation
- One South Asian form
- One Chinese form
- One form from New Guinea
- One form from Australia
- One American form

These "forms" of rice are a result of millions of years of geographic isolation and separate cultivation. Botanical evidence for the origin of rice is based on where wild rice relatives are distributed geographically and from which it is surmised that current rice cultivars have been derived. A diversity of wild rice species is found in monsoon regions of eastern India, Thailand, Laos, northern Vietnam, and southern China, lending support to the theory that one or more of these areas served as the cradle of rice domestication. Archeological evidence for the dietary importance of rice in society dates back as early as 2500 B.C. in Mohenjo-Daro and the Yangtze basin in the late Neolithic era. Rice cultivation in the New World began in approximately 1685 by early settlers on the coastal lowlands and in the South Carolina region. In the 18th century, rice production spread to Louisiana and in the 20th century to the Sacramento Valley of California. A rice genomic project is currently under way at the Institute for Genomic Research (TIGR).

6.6 Case History: Evolution of Corn (*Zea mays*)

Since the 1920s, plant breeders and scientists have debated the origins of cultivated corn. The current hypothesis is that corn was domesticated approximately 9000 years ago in the Balsas River Basin, Mexico. Domestication occurred through a single event involving the wild relative *Teosinte parviglumis.* When comparing the genomes of *Teosinte* and *Zea,* five regions are responsible for the phenotypic difference between the two genera. Several genes have been identified and molecularly cloned, including tb1, pbf, and su1. These genes encode for corn kernel proteins and starch debranching enzymes. When comparing these genes in modern maize varieties with ancient archeological specimens, these genes were found to have been conserved. This suggests that the genetic changes responsible for the attributes of modern corn compared to the ancient wild relative of *Teosinte* have been maintained for millennia. In addition, comparing Mexican corn samples from 44,000 years ago with modern corn showed these genes were comparable. This indicates that intensive and long-term human selective pressure on *Zea mays* varieties has resulted in a homogenization of the corn genome long before genetic modification by recently developed biotechnologies.

6.7 Conservation Biology

Conservation biology is an area of research that assesses levels of genetic diversity within a population for a given species. High levels of inbreeding (reproduction between close relatives) reduce genetic diversity within a population. Mass disasters and significant migration patterns can lead to genetic "bottlenecks" and a "founder effect" within populations. These terms refer to events that lead to a reduction in population size to a few individuals that become the founders of a new population. The genetic result of this situation is that the population is closely related and highly inbred, resulting in genetic homogeneity. Using historical records and molecular markers, current populations can be studied to determine the levels of genetic diversity. In particular, this may be an important consideration for organisms that are heavily poached (animals, fish) or heavily selected plant breeding seed sources. For organisms determined to have low levels of genetic diversity within their populations, forced-out breeding between the most genetically distinct individuals can alleviate some of the inbreeding effects (e.g., as for captive animal zoo stocks). For plants, either private commercial ventures or public societies maintain seed banks. The purpose of seed banking is to maintain current seed varieties, older "heirloom" varieties, and wild relatives. Every 3 years, stored seeds need to be grown into plants and their seeds harvested in order to maintain seed viability. In this manner, the existing genetic diversity for a plant species and its wild relatives can be maintained for future generations.

6.8 The Molecular Measure of Genetic Diversity

Short sequence repeat (SSR) loci have been identified in all eukaryotic and some prokaryotic species. Variations in the number of repeat sequences may result from normal genetic recombination processes or "mistakes" in genetic repair enzymes, such as slipped-strand mispairing by enzymatic polymerases. These markers also can be used to distinguish between individuals within a population, but not all markers will perform equally in their ability to discriminate between individuals. Many examples exist for the application of SSR loci to taxonomic and population studies for genetic diversity. A few examples are presented in this section.

6.8.1 SSR Markers for the Fungal Endophyte, *Sphaeropsis sapinea*

S. sapinea is a fungal endophyte of the *Pinus* species that may cause disease after trees are exposed to stress. Four morphologically different types have been described within the natural range of this endophyte. Eleven polymorphic SSR markers have been identified and tested on 40 different isolates of *S. sapinea*. These markers were able to clearly distinguish between the different types, and one type was found to be quite genetically diverse. All of the isolates could be further correlated with their geographic origin, which may prove helpful for biogeosourcing studies. In forensics, this type of information may be useful for tracing the origin of a wood sample obtained at a crime scene, for example. In addition, other types of fungal associations may be useful for both human and botanical forensic and taxonomic purposes.

6.8.2 Microsatellites in the Cultivated Peanut

Very few polymorphic markers have been identified for the cultivated peanut (*Arachis hypogaea* L.). To characterize different cultivars, SSR loci were identified using a standard microsatellite enrichment procedure. A GA/CT repeat was frequently dispersed throughout the peanut genome. Nineteen out of 56 different microsatellites showed genetic polymorphism when a sample population was surveyed. SSR loci, thus, show some potential for measuring genetic diversity within peanut cultivars.

6.8.3 The Use of Microsatellites to Study Inter- and Intraspecific Genetic Variation

A common SSR or microsatellite used in DNA fingerprinting is the GATA repeat sequence. A Basmati rice variety of *Oryza sativa* was used for the cloning and characterization of three polymorphic GATA repeats. These markers were then used to survey 26 wild relatives of rice, 16 cultivated rice genomes, 47 elite Indian rice varieties, and 37 rice lines that were known to be either bacterial blight disease susceptible or resistant. Up to 22 different alleles were observed for one of the marker loci showing a high degree of discrimination power. These markers and others should be useful for characterization of levels of genetic diversity within a crop that has been extensively cultivated by humans.

6.8.4 Molecular Markers in Tree Species Used for Wood

Wood is used for many purposes in human society and, therefore, may be present at both indoor and outdoor crime scenes. It is used for fuel, shelter, furniture, and heat, to name just a few critical applications. If dry wood were amenable to molecular genetic investigations, this could lead to a major advance in forensics. To evaluate the potential of wood for molecular genetic investigations, researchers have attempted to isolate and amplify DNA fragments corresponding to three plant genomes from different regions of ten oak logs. Stringent procedures to avoid contamination with external DNA were used in order to demonstrate that the fragments were from oak and not an additional fungal or mold source on the wood. This authenticity was further confirmed by demonstrating genetic uniformity within each log using both nuclear and chloroplast microsatellite markers. For most wood samples, the DNA was degraded, and the sequences that gave the best results were those of small size and present in high copy number. Overall, this work demonstrates that molecular markers can be used for genetic analysis on dry oak wood, but there is a need for further evaluation of the potential of wood for DNA analysis.

Accurate wood identification depends on the microscopic characteristics of the wood cells. As a first step, a cross-sectional slice of the sample is examined with a hand lens. Then thin sections are cut freehand along the grain from the radial and tangential surfaces and prepared for microscopic analysis and species comparison and identification. Features like color, odor, and texture are of limited identification value as they are quite variable. Chemical features are sometimes employed in the routine identification of wood. The use of DNA technology, once thought to be impossible for dry wood, is now feasible, though in its infancy. Modern techniques have allowed extraction from dry material, and it is reasonable to believe that the next decade will see the entrance of DNA markers for accurate and specific identification of commercially important species of wood.

In fact, microsatellite DNA/simple-sequence-repeat (SSR) loci were recently identified, isolated, and characterized in white spruce (*Picea glauca*). Inheritance and linkage of polymorphic SSR loci were determined in F1 progeny of four controlled crosses. The new markers were also assessed for compatibility and usefulness in five other *Picea* species. DNA sequencing of 32 clones selected from a molecular screen of two libraries identified 24 microsatellite markers. Eight microsatellite DNA loci were of the single-copy type, and six of these were polymorphic. A total of 87 alleles were detected at 6 polymorphic SSR loci in the 32 *P. glauca* individuals drawn from several populations. These microsatellite DNA markers, when developed, could be used for assisting various breeding, biotechnology, tree forensics, conservation, restoration, and sustainable forest management programs in spruce species.

6.9 Summary

Perhaps inherent to human nature is the need to catalog, organize, and classify all objects and organisms we encounter. Many organisms group easily together, whereas others are difficult to classify due to inconsistent morphology, conflicting phenotypic traits and molecular markers, or lack of variability in definable traits or markers. While lists of common plant species and the use of phylogenetics in classification can be handy as references, they also serve another function. Morphological and molecular markers are both extremely useful for comparing class characteristics for forensically relevant evidence. By studying the basics of plant diversity and evolutionary biology, forensic science "borrows" from these areas of scientific study in order to perform appropriate analyses of physical traits of trace plant evidence.

General References

Bishop, C.L., Ancient "GM" corn, *The Scientist*, reference available at www.the-scientist.com, 2003.

Deguilloux, M.F., Pemonge, M.H., and Petit, R.J., Novel perspectives in wood certification and forensics: dry wood as a source of DNA, *Proc. R. Soc. Lond. B. Biol. Sci.*, 22, 1039, 2002.

Doebley, J., Mapping the genes that made maize, *Trends in Genetics*, 8, 302, 1992.

Hagmann, M., A paternity case for wine lovers, *Science*, 285, 1470, 1999.

Huke, R.E. and Huke, E.H., Rice: then and now, reference available at www.riceweb.org, 1990.

Jaenicke-Despres, V. et al., Early allelic selection in maize as revealed by ancient DNA, *Science*, 302, 1206, 2003.

Matsuoka, Y. et al., A single domestication for maize shown by multilocus microsatellite genotyping, *PNAS*, 99, 6080, 2002.

Rajora, O.P. et al., Isolation, characterization, inheritance and linkage of microsatellite DNA markers in white spruce (*Picea glauca*) and their usefulness in other spruce species, *Mol. Gen. Genet.*, 264, 871, 2001.

The Use of Biological and Botanical Evidence in Criminal Investigations

7

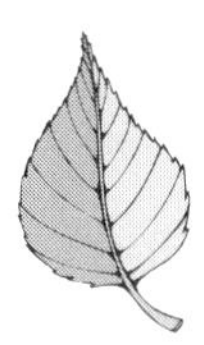

CARLL LADD AND HENRY C. LEE

Contents

0-8493-1529-8/05/$0.00+$1.50

7.1 Introduction

Forensic science is the application of science to matters of law. This general definition broadly encompasses many different scientific disciplines from which forensic science can acquire scientific testing procedures, including chemistry, biology, botany, entomology, physics, and the material sciences. The forensic sciences include crime scene reconstruction, evidence processing, biology (e.g., presumptive and confirmatory tests for semen, saliva, and blood), firearms/ballistics, latent prints, document examination, forensic photography, chemistry (e.g., arson and explosives), trace physical evidence (e.g., paint, hair, and fiber analyses), computer crimes, and toxicology. Within each of these broad categories of the forensic sciences are areas of specialization. For example, within biology a forensic scientist could specialize in DNA analysis and further specialize in a specific area of DNA (e.g., mitochondrial DNA analysis). Forensic botany could be considered a broad category of forensic science or a specialty area within biological or trace physical evidence depending on the extent of forensic botanical examination and the structure of the forensic testing laboratory. Some commonly accepted forensic specialties are forensic botany (legal applications of plant science), forensic medicine (medical jurisprudence), forensic odontology (dental application to human identification), forensic anthropology (identification of skeletal remains), forensic toxicology (identification of toxic substances in human tissues and organs), forensic entomology (knowledge of insect habitat and life cycles to approximate time of death), forensic archeology (examination of burial sites and past crime scene reconstruction), forensic engineering (including traffic and airplane investigation, mechanical failure, etc.), and forensic meteorology (application of weather conditions to forensic issues).

Crime scene investigation is based on the principles of transfer theory, in particular, Locard's Exchange Principle. Locard's Exchange Principle states that any time two or more surfaces come in contact with each other, there is a mutual exchange of trace matter between the two surfaces. A common generalization from this principle is "every contact leaves a trace." This concept is critical for considering how evidentiary samples can be utilized to link a suspect to a victim, a victim to a crime scene, or a suspect to a scene. Crime scene investigation can be further considered as the first and most critical step of any forensic investigation. If evidence is not recognized or is improperly collected or packaged, it may be inadmissible in court. For all physical evidence, including botanical samples, there are three crucial steps that must be adhered to at a crime scene: recognition of the evidence, appropriate collection methods, and adequate preservation of the evidence for later testing and use at trial. Once the physical evidence has been collected at a crime scene, it is delivered to a forensic testing laboratory where identification of the evidence and further individualization of evidentiary samples is performed. The identification of evidence means classification of a sample into a physical, biological, or chemical category. Depending on the assigned category, identification of a sample (e.g., hair) would place it into a biological grouping where a microscopic comparison to known reference samples would be made to determine if a person could be a possible contributor of the biological hair sample. The microscopic comparison to known reference samples in the case of hair examination would be useful but not considered an individualizing test. To individualize a sample, DNA analysis (STR typing) would then be performed on any epithelial cells from the hair root. Sample identification followed by individualization, when possible, is a general practice in all areas of forensic science regardless of testing specialization. In addition, pattern recognition and comparison to known reference samples is a common forensic theme that applies to DNA, latent prints, document examination, and so forth.

During the past few decades, the courts have placed greater reliance on objective scientific evidence. Eyewitness accounts can be unreliable or biased. However, forensic evidence such as DNA, fingerprints, and trace evidence (including botanical evidence) may independently and objectively link a suspect/victim to a crime, disprove an alibi, or develop important investigative leads. Physical evidence has also demonstrated its worth as an effective tool for exonerating those wrongfully accused of crimes. Biological evidence, including plant material, has been associated with many types of crime, but is typically associated with violent crimes such as homicide, physical assault, sexual assault, and motor vehicle accidents ("hit and run" cases). The natural consequence of its greater emphasis is increased legal scrutiny. Evidence integrity begins with the first investigator at the crime scene. Celebrated court cases such as the O.J. Simpson and JonBenet Ramsey homicides highlight standard challenges to the use of biological evidence in criminal investigations and suggest that the scrutiny of evidence collection, preservation, and handling will continue unabated. The entire case may be jeopardized if evidence is mishandled during the initial stages of the investigation. Indeed, evidence that is not properly recognized, documented, collected, and preserved may ultimately be of no probative value. This chapter addresses the issues surrounding the recognition, collection, preservation, and analysis of biological evidence (in general) and the emerging use of botanical evidence in criminal investigations. The general concepts for each major heading are presented first, and then, where appropriate, the special considerations for botanical evidence are considered.

7.2 Physical Evidence and Crime Scenes

Physical evidence is often classified as biological, chemical, or pattern evidence. Physical evidence can also be divided further into five general categories: transient, conditional, pattern, transfer, and associative evidence (see Table 7.1). Another useful way to view physical evidence is related to the type of crime scene. Crime scenes can be thought of as at least two distinct entities. First, there is the global or macro crime scene, which consists of the location of the crime, the victim's and suspect's body, and the surrounding area. Second, the body can be considered a micro crime scene by itself and could contain botanical material or transferred plant stains, critical body cavity evidence (e.g., oral, vaginal, and anal), or other trace material under the fingernails. Furthermore, the condition (is the scene organized or disorganized?) and nature of the crime scene (is the scene active or passive?) are also important elements to the investigation. Physical evidence can be used to gain information on the corpus delicti, the modus operandi, linkages between persons and crime scenes, witness statements/testimony, investigative

Table 7.1 Types of Physical Evidence

Transient	Evidence easily lost or changed over time, e.g., any blood or body fluid stains subjected to the elements.	Body temperatures, odors, bite marks, tool marks, and degree of decomposition.
Conditional	The result of an action or event. Can be transient in nature — lost unless properly documented.	Lividity, rigor, stomach contents (animal and plant), weather conditions, lighting, insects.
Pattern	Markings, imprints, indentations, or other patterns found at the scene.	Tire and shoe impressions, wound patterns, glass fractures, bite marks, blood spatter, incendiary patterns, weapon patterns, projectile trajectories.
Transfer	The physical exchange of material when objects come in contact.	Paint chips, hairs, fibers, soil, glass, plant matter.
Associative	Evidence that can link a victim or suspect to a scene or each other.	Wallets, identification cards, key, jewelry, photos, clothing, money.

Table 7.2 Sources of Biological Evidence

Body fluids	Blood, semen, saliva, urine, tears, bile, perspiration
Soft tissue	Skin, organs, hairs, fingernails
Structural tissue	Teeth, bones, skull
Other human	Fecal matter, vomit
Objects contacted by persons	Lipstick, tooth brush, hairbrush, electric razor, cigarette butts, drinking glasses, articles of clothing
Nonhuman	Microbial, animal, plant (matter and stains)

leads, suspect identification, substance identification, and ultimately crime scene reconstruction.

7.3 Sources of Biological and Botanical Evidence

Biological evidence originates from either the victim, the suspect, or persons other than the victim. Fifteen major sources of biological evidence are associated with violent crimes (see Table 7.2). Sources of botanical evidence can include pollen, leaves, petioles, roots, stems, fruits, and seeds. These types of evidence can be found at all crime scenes but may be particularly informative for outdoor crime scenes. Plant matter can be from stomach contents, wood from a weapon or from a tree embedded in the undercarriage of an automobile, fecal matter, stains on clothing, vegetation covering bodies, or toxic material in a deadly overdose or related to drug trafficking.

7.4 Transfer of Biological and Botanical Evidence

Biological evidence can be used to make linkages or associations (e.g., person–person, person–other physical evidence, person-crime scene). In general, biological evidence can be transferred by direct deposit or by secondary transfer.

7.4.1 Direct Deposit

Any biological evidence (blood, semen, body tissue, bone, hair, urine, and saliva) can be transferred to an individual's body/clothing, object, or crime scene by direct deposit (primary transfer). Once biological fluids are deposited, they adhere to the surface and become stains. Nonfluid biological evidence, such as tissue or hair, can also be transferred by direct contact. Common examples of the direct deposit of botanical evidence include seeds being transferred to clothing (especially trouser cuffs, pockets, and hems), stains from direct pressure (such as might occur during an assault), or pollen on shoes after brushing against a flowering shrub.

7.4.2 Secondary Transfer

Blood, semen, tissue, hair, saliva, or urine can be transferred to a person, object, or location through an intermediary (person or an object). With secondary transfer, there is no direct contact between the original source (donor of the biological evidence) and the target surface. Secondary transfer may establish a direct link between an individual and a crime. For botanical evidence, this type of transfer may be an important factor in a subset of cases. Secondary transfer of botanical evidence may be possible with some wind-dispersed pollen, seeds, clinging vines, thorns, and burrs. Animal and bird intermediaries may also play a significant role in the secondary transfer of botanical evidence for seeds and burrs.

The impact of secondary transfer on the interpretation of DNA results has been debated. However, clearly the specter of secondary transfer is more likely to be raised in cases that involve the more sensitive DNA typing methods such as mitochondrial DNA sequencing and low copy number polymerase chain reaction (PCR). The secondary transfer from shed epithelial cells (from one object to another through an intermediary) has raised some concern over collecting DNA samples from objects that are handled by multiple people (e.g., doorknobs, telephone receivers). Few studies have been performed regarding the secondary transfer of botanical evidence and plant DNA typing; however, common principles should apply to this form of evidence as well.

7.5 Evidence Recognition

A pivotal step in processing a crime scene is recognizing which samples warrant further testing. This phase may be challenging, as crime scenes can be both complex and chaotic. Hence, an experienced investigator who systematically evaluates the scene is an invaluable resource. Recognition is the ability to identify probative evidence (at the scene or in the lab) scattered among potentially vast quantities of redundant, irrelevant, or unrelated items. For instance, collecting 20 bloodstains from the vicinity of a stabbing victim may not point to the perpetrator. The recognition process involves basic forensic principles such as pattern recognition and analysis. If crucial evidence is not recognized, collected, and preserved, its value will be lost. Many outdoor crime scenes have trace plant evidence that is overlooked by crime scene personnel. Plant evidence can be associated with a particular individual or provide other important investigative leads, for example, whether the crime scene is primary or secondary. In general, where the original crime takes place is the primary scene and any other subsequent scene is called a secondary scene.

7.5.1 Recognition of Botanical Evidence at Crime Scenes

Botanical evidence can be found associated with a wide variety of crimes. The exact location of the botanical evidence and the type of samples recovered will be highly dependent on the type of crime. Botanical evidence that appears foreign to the crime scene is the easiest type to recognize and occurs primarily as displaced leaves, seeds, or pollen that do not appear associated with plants in the vicinity. Some types of botanical evidence are more difficult to identify, such as leaves that have been removed from a plant or broken branches. It is useful to document this type of information, but one must bear in mind that a broken branch may not actually be associated with the crime. However, if leaves or seeds are discovered on a body in an abandoned automobile, it may help in determining the primary crime scene and the perpetrator.

Primary crime scenes are where a body or criminal event has occurred and the evidence and the body remain at the site. Secondary crime scenes are where, for example, the body was dumped but not where the homicide took place. Another example of a secondary crime scene is where stolen equipment is recovered in a toolshed but was removed from the primary residence several miles away. At a primary crime scene, botanical evidence can be located in, on, or around the body. Some examples include leaves on the back of a victim, algae in the mouth of a victim, or pollen grains inhaled by the victim and lodged in the nasal cavity. At a secondary crime scene, leaves, branches, and pollen associated with a body, caught in the undercarriage of an automobile, or retained inside stolen electronic equipment can provide a link back to the primary scene. Botanical evidence may be useful in determining or reconstructing key aspects of the crime. For example, all crime scenes have the following elements: a point of entry, a path traveled by the perpetrator, a target area where the crime occurred, and a point of exit (see Table 7.3).

7.6 Documentation of Biological and Botanical Evidence

The location and condition of any biological evidence must be thoroughly documented prior to its collection. Careful evidence documentation is essential both at the crime scene and the forensic laboratory. In any criminal or civil investigation, documentation has great bearing on whether the evidence can later be introduced in court. Evidence should not be processed or moved

Table 7.3 Elements of Crime Scenes

Elements of Crime Scenes	Evidence
1. Point of entry	Damaged shrubbery, patterns of footwear, tire tracks or other imprint evidence, compressed grass where a perpetrator may have stood or an object may have been dragged — pollen composition may be important.
2. Path traveled by perpetrator	Broken branches, compressed grass or underbrush, wood chips left on a carpet that match those at the point of entry, paints, body fluids or other stains, prints, clothing, personal effects.
3. Evidence on or in the victim	Leaves, pollen, algae, plant/clothing fibers, marijuana or other drugs, tobacco or other cigarette components (e.g., paper, wooden matches), food stains, or stomach contents. Be aware that the surface of the body, all body orifices, under the body, and in the hands or under the fingernails should be searched for evidence.
4. Point of exit	May include all of the items mentioned for point of entry (#1), as well as latent prints, bloodstains, soil, dust, and glass, which may be useful as physical evidence.

until its original condition and other relevant information have been recorded. Several means of documentation are available, and generally, the use of more than one method is advised. In order, the following progression of steps is advised: notes, photography, videotaping, sketches and diagrams, and audiotaping. Note-taking is a simple method for reminding an investigator of minor details that may be important later for official records and reports. Information that can be included as written notes should be dates, times, a brief description of the crime scene location, method of notification, persons noted at the crime scene, the officer in charge, lighting, weather conditions, odors, and noise.

For botanical evidence, an investigator should take note of the location of any flowering plants and shrubs (flower odor, possible pollen source), broken fragments of wood (loose or from doorsills and windowsills), broken tree branches or stems, trampled areas containing plants, and any plants bearing seeds or fruit. In addition, any food items that may be found at the crime scene should be noted, as they may be useful later for establishing or refuting an alibi or for determining time of death. In particular, food with bite marks should be carefully considered and recorded. Photography, video, and audiotaping along with simple diagrams can be useful later for crime scene reconstruction. In particular, a sketch of the outdoor landscaping at a crime scene and weather conditions can be crucial for establishing botanical linkages especially as some plant species may be considered transient evidence based on growth seasons. For example, many spring blooming plants are only visible for the first few weeks in March/April in the northeastern U.S.

7.7 Collection and Preservation of Biological and Botanical Evidence

7.7.1 General Biological Samples

Today, the primary tool for individualizing biological evidence is DNA analysis. However, one of the more effective challenges to DNA involves attacking the integrity of the sample before it arrives at the forensic laboratory. Indeed, the ability to introduce DNA findings in court has been adversely affected by challenging evidence collection and preservation procedures. Detailed evidence collection protocols have been well described in a variety of textbooks. The specific collection method employed will depend on the state and condition of the biological evidence. In general, an investigator should collect a significant quantity of material to ensure recovery of sufficient DNA for testing purposes. However, it is important to limit collecting additional dirt, grease, fluids, etc. from the surrounding area since many substances are known to adversely affect forensic analysis methods such as the DNA typing process. Carefully collect each item to minimize contamination using disposable gloves, forceps, sterile swabs and scalpels, particle masks, and so forth as appropriate for the sample. Package each item of evidence separately according to established forensic practices to minimize cross-contamination. The evidentiary value may be seriously compromised if several items are packaged together. In addition, biological samples such as wet bloodstains should be packaged in sealed paper bags, not in airtight containers, which retain moisture and promote bacterial growth. It is useful to place any moistened swabs in a specially designed cardboard swab collection box that facilitates sample drying during transport. Once the samples have been collected and packaged, they should be promptly delivered to the forensic laboratory. To minimize specimen deterioration, items should be stored in a cool, dry environment until they are submitted for testing.

7.7.2 Botanical Samples

The collection of plant fragments, seeds, flowers, and fruits should all be performed by hand. Whole plants and any fragments that may potentially be useful for a physical match should be collected as well as any pieces associated with a body. Botanical fragments in and on motor vehicles should be collected; in particular, the wheel wells, in and under floor mats, the undercarriage, pedals, windshield wipers, vents, trunk, and engine compartments should be fully examined. Care should be taken to avoid vacuuming a vehicle for microscopic trace evidence until all plant fragments have been collected by hand. Botanical evidence should be collected in paper, not plastic, when possible. Plant matter in stomach contents would most likely be collected during autopsy by the medical examiner. Fecal matter deserves a special note on collection methods since it differs from most aspects of botanical collection. For fecal matter and later possible plant identification, the sample should be kept moist. If a fresh deposit, then the sample should be stored in glass or plastic. Dried feces or soiled clothing can be placed in paper bags and sealed. For microscopic analysis for plant cell identification, a 10% formalin solution as a preservative is acceptable. However, if plant DNA analysis may be performed, storing the sample in a freezer is optimal.

7.7.3 Outdoor Crime Scene Considerations

Outdoor crime scenes also warrant a special note. Investigator and technicians are often working under greater time pressure because, given certain weather conditions, some of the physical evidence can be altered or destroyed. Animals and insects can change the condition of the body, and the rate of sample decomposition can be great. The size or scope of the scene may be large or somewhat undefined and hence challenging to secure. The media and the general public can interfere or impede the collection process. A complete record of all crime scene personnel is essential. Elimination fingerprints, shoeprints, and DNA samples may need to be collected from scene technicians and other individuals to understand the significance of the physical evidence. For example, a cigarette butt recovered near the location of the crime could be used for DNA analysis. If the evidentiary DNA profile excludes the suspect, it could point to a different perpetrator. However, if a neighbor or a police officer was the source of a cigarette butt or shoeprint at the scene, a different conclusion could be more valid. Hence, it is vital that access to the scene be restricted to persons associated with the investigation.

7.7.4 Examination of a Body

The body of the victim/decedent must be thoroughly examined and documented prior to sample collection. Describe and sketch/photograph the location of any possible injury, trace evidence, or stains on the victim. Note potential bite marks, bruises, abrasions, and so forth. All trace evidence, including blood and body fluids, prints, soil, hairs and fibers, pollen, plant fragments, and stains, should be photographed in place prior to collection and packaging. Body orifices should be examined for semen and other body fluids, hairs and fibers, and other trace materials. It is important to collect the evidence when possible prior to moving the body. Physical evidence has been lost or altered when the body is placed in a body bag and the evidence is then washed away at the autopsy room. In addition, to minimize the recovery of skin cells from the victim, collect any blood/body fluid stains from the putative suspect as gently as possible. This can be accomplished by lightly swabbing the stained area. Pubic combings are regularly collected to identify evidence that could have been transferred from the perpetrator. Fingernail scrapings or clippings are often collected in homicides and sexual assaults. Here, too, it is important to avoid applying excessive force in this process to minimize collecting the victim's blood/skin. In general, collecting too much biological material from the victim may mask or impede the detection of foreign biological evidence, that is, from the perpetrator. It is also important to document (sketch and photograph) any pattern evidence on the victim's clothing, such as blood spatter, smears, and swipes. Carefully collect the victim's clothing to minimize the loss of important trace evidence and preserve any pattern evidence. It is often useful to have the victim disrobe over a large piece of clean paper. If clothing must be cut from the victim, avoid cutting or damaging any possible bullet/knife holes.

7.8 Sources of DNA

7.8.1 General DNA Typing Considerations

Many different types of biological evidence are commonly submitted to forensic science laboratories for examination. Initially, evidence that was suitable for DNA analysis was limited to human biological substances that contain nucleated cells. This limitation has been overcome in the last 5 years with the implementation of mitochondrial DNA sequencing, and plant and animal DNA testing in the forensic arena.

The quantity of DNA that can be extracted from common biological sources varies greatly with evidentiary samples. Note that, in practice, crime scenes samples may contain considerably less usable DNA depending on environmental conditions. DNA has been isolated from other sources such as gastric fluids and fecal stains. However, it can be difficult to generate a DNA profile from these sources due to significant degradation.

Four factors affect the ability to obtain a DNA profile. The first issue is sample quantity. The sensitivity of PCR-based DNA typing methods is noteworthy but still limited. The second concern is sample degradation. Prolonged exposure of even a large bloodstain to the environment or to bacterial contamination can degrade the DNA and render it unsuitable for further analysis. The third consideration is sample purity. While most DNA typing methods are robust, dirt, grease, some dyes in fabrics, and so forth can seriously compromise the DNA typing process. Environmental insults will not change DNA type "A" into type "B," but they can adversely affect the ability of the scientist to obtain a complete DNA profile from the sample. The last issue is the ratio of major contributor to the minor contributor(s) from DNA mixtures. Both DNA profiles can readily be detected with, for example, 1:1 and 3:1 DNA mixtures. However, with mixture ratios such as 25:1 or 50:1, the quantity of the major DNA species prevents the detection of the minor source.

7.8.2 Plant DNA Typing Considerations

For DNA analysis of botanical samples, many of the same considerations for human identity testing still apply. The four factors (quantity, quality, purity, and mixture ratios) may all play a role in determining whether further testing should be performed or whether an interpretable DNA profile can be obtained from a plant sample. Two factors need to be considered for obtaining a sufficient quantity of plant DNA for profiling: size of the plant fragment and ability of the analyst to mechanically break the plant cell wall for the sufficient release of nuclear DNA contents. The size of a plant fragment can seldom be enhanced other than to initially collect as much sample as possible from the crime scene. However, new technologies are aiding in the DNA extraction process for improving DNA yield from plant samples. Traditionally, plant cells have been disrupted by the mechanical pressure of grinding by hand in a mortar and pestle with the addition of liquid nitrogen to increase the fragility of the cell wall. Some useful equipment for breaking plant cell walls may be a mechanical homogenizer (consisting of a rotating blade with a serrated tip) or the addition of metallic beads prior to high-speed oscillation of samples. In addition, several companies manufacture commercial plant DNA extraction kits that minimize the number of centrifugation steps and maximize DNA yields. The use of some of these commercial kits also seems to improve the final purity of the plant DNA so that the PCR amplification process is not inhibited by the presence of secondary plant metabolites (e.g., tannins, resins, phenolic compounds). Finally, care in collection and preservation of botanical evidence as well as the development of plant species-specific probes and PCR primer sets in the future will address mixture interpretation issues.

7.9 Preparation of Genomic DNA from Plant Tissue

Although many commercially available plant DNA extraction kits have recently become available, there are some very reliable traditional chemical extractions for plant DNA purification. The following protocol is one example of such a procedure.

Materials:

Cold, sterile water
Liquid nitrogen
Extraction buffer (100 m*M* Tris-Cl, pH 8; 100 m*M* EDTA; 250 m*M* NaCl; 100 µg/mL proteinase K)
10% (wt/vol) N-lauroylsarcosine
Isopropanol
TE buffer (10 m*M* Tris-Cl, pH 8; 1 m*M* EDTA)
Cesium chloride (CsCl)
10 mg/mL ethidium bromide
CsCL-saturated isopropanol
Ethanol
3 *M* sodium acetate, pH 5.2
Beckman JA-14, JA-20, JA-21, and Vti80 rotors

Harvest 10–50 g of fresh plant tissue.
Rinse with cold, sterile water; dry with tissues; and freeze with liquid nitrogen. Mechanically grind with a mortar and pestle.
Transfer frozen powder to a 250-mL centrifuge bottle and immediately add 5 mL of extraction buffer per gram of starting fresh plant tissue; gently mix. Add 10% N-lauroylsarcosine to a final concentration of 1%; incubate 2 h at 55°C.
Centrifuge for 10 min in a JA-14 rotor at 6000 rpm chilled to 4°C. Save the supernatant and repeat this step if debris is still present.
Add 0.6 volumes of isopropanol and mix. If no visible precipitate forms, place at –20°C for 30 min. Centrifuge for 15 min in a JA-14 rotor at 8000 rpm at 4°C. Discard the supernatant.
Resuspend the pellet in 9 mL TE buffer, add 9.7 g of solid CsCl, mix, and incubate 30 min on ice. Centrifuge for 10 min in a JA-20 rotor at 8000 rpm at 4°C and save the supernatant.
Add 0.5 mL of 10 mg/mL ethidium bromide and incubate 30 min on ice. Centrifuge 10 min in a JA-20 rotor at 8000 rpm at 4°C. Transfer supernatant to two 5-mL ultracentrifuge tubes and seal.
Centrifuge in a Vti80 rotor at 20°C for 4 h at 80,000 rpm. Collect the band of DNA using a 15-G needle and syringe.
Remove any residual ethidium bromide by repeated extraction of the DNA band with CsCl-saturated isopropanol.
Add 2 volumes of water and 6 volumes of 100% ethanol, mix, and incubate for 1 h at –20°C. Centrifuge for 10 min in a JA-20 or JA-21 rotor at 8000 rpm at 4°C.

Resuspend the pellet in TE buffer and precipitate again by adding 1/10 volume of 3 *M* sodium acetate and 2 volumes of 100% ethanol. Repeat the centrifugation step. Air-dry the pellet by inverting the tube over a paper towel for 15 min, then resuspend the pellet in the TE buffer. The expected plant DNA yield should be 104 μg of 50 Kb high-molecular-weight DNA per gram of starting plant tissue.

7.10 Case Examples of Plants and Homicide

7.10.1 The Wood Chipper Case: Linkage of a Vehicle to a Crime Scene through Wood Fragments

When Helle Crafts disappeared, her husband, Richard Crafts, told police that she left her children in Newtown, CT, and flew home to Denmark to care for her ailing mother. Richard Crafts' story began to unravel, however, as concerned neighbors and friends insisted that something more sinister had happened to Helle. Helle took no clothes with her, and her mother was found to be quite healthy and unaware of Helle's whereabouts. It was also determined that Richard Crafts rented a hook for towing heavy equipment, reserved a Badger Brush Bandit wood chipper, and purchased a chainsaw and a Westinghouse chest freezer — all within days of Helle's disappearance. A few days later, a snowplow driver reported that he could not complete his route because a wood chipper was left at the side of the road by the Housatonic River. Investigators went to the area where the wood chipper was last seen and recovered physical evidence that ultimately led to the conviction of Richard Craft for the murder of his wife. This evidence included a tooth, a partial fingernail, some bone fragments, wood-chips, tissue, and approximately 4000 strands of human head hair. Unfortunately, the DNA tests available in 1986 for forensic casework were less sensitive than current methods, and the DNA test results were inconclusive. However, plant evidence did play a significant role in the case. The bone fragments and wood chips recovered from the riverbank along with the woodchips from the rental truck were shown to have the same multiplane cut pattern, consistent with the cut of the wood chipper (see Figure 7.1, Figure 7.2, and Figure 7.3).

7.10.2 Of Leaves and Men: The Identification of a Body through Leaf Litter Composition

KRISTINA A. SCHIERENBECK

On Tuesday, June 4, 2002, two criminalists from the California Department of Justice called the Department of Biological Sciences at California State University in Chico, CA. They were looking for a botanist with regional, taxonomic, and ecological expertise in Northern California. They were at a

Figure 7.1 Depicted are representative wood chips from the Helle Craft crime scene where body disposal occurred.

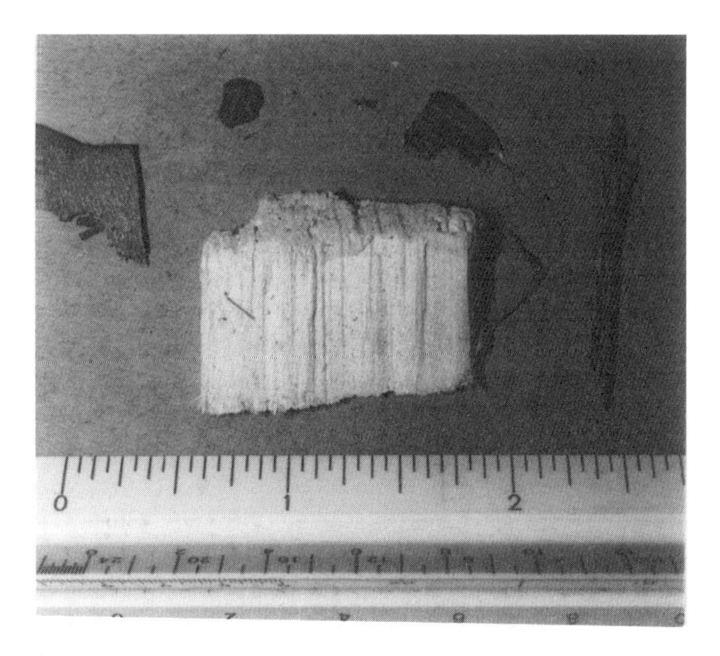

Figure 7.2 The plane markings on some of the wood chips matched the wood chipper in question.

Figure 7.3 The Brush Bandit wood chipper that was used for disposal of Helle Craft's remains after her murder, as shown here.

dead end in a case of a missing child. They had found the father of the child deceased from a self-inflicted gunshot wound in his truck along a mountainous road (Highway 32) in Butte County, CA. The father and daughter had last been seen in Las Vegas, NV, and the only clue was some vegetation found in the truck. The criminalists provided leaf litter collected from clothing of the missing girl. The clothing and associated leaf material (approximately 2 grams) were matted with blood. Based on leaf composition that ranged from whole to partially decomposed leaves, the sample was determined to be from the top centimeter of leaf litter. The sample contained an intact fresh leaf of greenleaf manzanita (*Arctostaphylos patula*) torn from a living shrub. Fragments of other plant species were also present (in order of abundance): canyon live oak (*Quercus chrysolepis*) or interior live oak (*Q. wislizenii* var. *wislizenii*), white fir (*Abies concolor*), greenleaf manzanita, ponderosa pine (*Pinus ponderosa*), and black oak (*Quercus kelloggii*). A preliminary identification of these species was made, and then they were compared to known reference samples from the California State University, Chico, Herbarium. Each sample was examined for leaf venation characteristics, leaf margin characteristics, and general appearance. A positive identification for the oak species was not possible. The leaves were identified as either canyon or interior live oak based on the dried samples; oak species are notoriously variable and readily hybridize with one another.

Based on known species distributions and ecological site requirements, certain geographic regions could be eliminated as locations for the missing body. The species identified do not occur together in Nevada or on the eastern exposure of the Sierra Nevada. The live oak, in particular, indicated that the sample was most likely on the western exposure of the Sierra Nevada. Overlapping species distributions indicated the elevation of the sample to be from an elevation of 980 to 1380 m. The relative abundance of the species present was a bit unusual in that the dominants were live oak and white fir and indicated that the site had to have both slightly mesic (white fir) and slightly xeric (live oak) characteristics. It was unlikely that the site was on a north- or east-facing slope due to the presence of the live oak, and a south-facing slope was equally unlikely based on the presence of the white fir. The leaf litter sample was predicted to be from a site on a west-facing slope with some available moisture. Further, the composition and dark color of the leaf litter sample indicated a high organic content, which placed the body site under a fairly dense forest canopy. The notable occurrence of the greenleaf manzanita indicated there had to be some available light at the site, despite the dense canopy.

Detective Borgman of the Butte County Sheriff's department called on Wednesday, June 5, 2002, to survey possible sites that afternoon. It was not until the survey reached the location of the truck that the species composition appeared to match that found in the leaf sample. A few miles past the truck location (the area already had been searched thoroughly), sites that had a dominance of live oak and white fir, a fairly closed canopy, with some greenleaf manzanita nearby, were noted. As the vegetation was surveyed at each site, Detective Borgman took specific location readings with a Geographic Positioning System. At the fifth site and close to the location where the truck was found, the litter met the outlined site criteria and a strong odor of decay and a blanket near a log was observed. Detective Borgman investigated further and uncovered the missing girl's body under the blanket. The location of the body between the logs had hidden the body from earlier searches.

How close were the site predictions based on leaf litter composition? The location of the body was on a west-facing slope with canyon live oak and white fir as the dominant species. A large canyon live oak was at the interface of the dirt highway turnout and a 30% west-facing slope was dominated by a closed canopy of coniferous vegetation. The location of the body was within three meters of a small patch of exposed chaparral with greenleaf manzanita as the dominant species. A tree species present at the site that was not identified in the leaf litter sample was Douglas fir (*Pseudotsuga menziesii*).

Based on the amount of blood found in the truck on the passenger side and the medical examiner's report, it appears the father shot his daughter point blank in the head. He removed her shirt to prevent blood flow (there was no blood at the location where the truck must have parked), and carried the body down the slope in the dark to hide it. As he wandered down the slope he crossed into the chaparral where the manzanita leaf was removed from the shrub. He placed her body on the forest floor between two felled logs, pulled the shirt from under her head where he picked up the leaf litter, and took the clothing back to the truck. It is unclear why he took the bloody clothing back to the truck. Perhaps as a message to her mother, he wanted to make clear that he had killed her but did not want the body to be discovered.

7.11 Summary

The role of a forensic scientist is to provide linkages or associations between people, crime scenes, and key objects. Physical evidence, when recognized, collected, preserved, and appropriately tested, can lend tremendous value to an investigation. The role of the investigator, scientist, and attorneys in the case is to sift through all of the evidence, examine all possible explanations, and provide some form of resolution to the case based on the evidence and results of forensic testing. Beyond the obvious linkages of suspect and victim to crime scenes, physical evidence can often be used to support or refute a witness statement, which can greatly clarify for investigators and jury alike what the events were that preceded the criminal act. Crime scene investigation is the first and most critical step in any case because if the evidence is not identified or collected properly, it may be inadmissible in court. In addition, most crime scenes are fluid in some fashion so that a crime scene today may not retain its appearance, weather conditions, odors, and so forth, making accurate and detailed documentation of the crime scene a key factor in the success of case resolution.

General References

De Forest, P.R., Gaensslen, R.E., and Lee, H.C., *Forensic Science: An Introduction to Criminalistics*, McGraw-Hill, New York, 1983.

Gaensslen, R.E., *Sourcebook in Forensic Serology, Immunology and Biochemistry*, U.S. Government Printing Office, Washington, D.C., 1983.

Herzog, A., *The Wood-Chipper Murder*, Henry Holt and Co., New York, 1989.

Lee, H.C., Coyle, H.M., and Ladd, C., DNA Typing Using Bone Samples in Human Identification Casework, in *To The Aleutians and Beyond*, Frohlich, B., Harper, A., and Gilberg, R., Eds., Department of Ethnography, National Museum of Denmark, Copenhagen, 2003, 227.

Lee, H.C. and Labriola, J., *Famous Crimes Revisited: From Sacco-Vanzetti to O.J. Simpson*, Strong Books, Southington, CT, 2001.

Lee, H.C. and Ladd, C., Preservation and collection of DNA evidence, *Croatian Med. J.*, 42, 225, 2001.

Lee, H.C., Palmbach, T.M., and Miller, M.T., *Henry Lee's Crime Scene Handbook*, Academic Press, Boston, 2001.

Lee, H.C., Ladd, C., Scherczinger, C.A., and Bourke, M.T., Forensic applications of DNA typing, collection and preservation of DNA evidence, *Am. J. of Forensic Med. and Path.*, 19, 10, 1998.

Lee, H.C., *Physical Evidence*, Magnani and McCormic, Enfield, 1995.

Lee, H.C., Gaensslen, R.E., Pagliaro, E.M., Mills, R.J., and Zercie, K.B., Physical Evidence in Criminal Investigation, Narcotic Enforcement Officers Association, Westbrook, 1991.

Lee, H.C., Identification and grouping of bloodstains, in *Forensic Science Review*, Saferstein, R., Ed., Prentice Hall, Englewood Cliffs, NJ, 1982, 267.

An Overview of Historical Developments in Forensic DNA Analysis

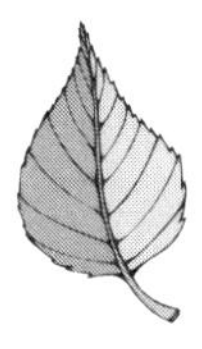

GARY SHUTLER

Contents

0-8493-1529-8/05/$0.00+$1.50

8.1 Protein Genetic Markers

The first genetic markers used in the analysis of biological evidence for criminal investigations were all protein based. They could be broadly characterized as serological markers, enzymes, and serum proteins. By far, the oldest and most widely used genetic marker in forensic analysis was the ABO typing system.[1] The predominance of serological-based methods then fit the term of *forensic serology* to describe the field of identifying and characterizing biological substances associated with crimes of violence. There are several comprehensive books that can be sought as references for this subject. In the U.S. at least two textbooks, *Forensic Science Handbook*[2] and *Sourcebook in Forensic Serology*,[3] are particularly notable. Throughout the 1970s and 1980s, many advances were made in the forensic application of other serological blood typing markers, serum protein, and red blood cell enzyme polymorphisms. These methods were widely used by crime laboratories to assist the police in successfully resolving criminal matters.

8.1.1 Analysis Limitations

Although very useful for excluding people from being the biological source of certain body fluid evidence, the number of possible contributors based on coincidentally shared types was high with serological techniques.[4] Even with results from several different blood typing systems, the probative value was such that a suspect could only be a possible source of the incriminating evidence. A bloodstain the size of a silver dollar was required for meaningful serological characterization. The more severe the environmental conditions the evidence was exposed to, the lower the chance of successful protein marker analysis. For evidence other than blood, such as semen and saliva, there were even fewer protein-based genetic markers that could be successfully applied.

8.2 The First Forensic DNA Analysis — RFLP Analysis

8.2.1 Jeffreys' Minisatellite Loci

In 1985 Alec Jeffreys, a researcher from the University of Leicester, published a series of papers in *Nature*, describing the use of core or consensus DNA sequences that could simultaneously detect many regions of DNA with highly variable lengths of tandemly repeated sequences.[5–7] The sites were called minisatellite loci and the process was named DNA fingerprinting. The nature of the process is based on Southern analysis (named after Ed Southern, who developed the gel-to-blot transfer method followed by subsequent hybridization[8]). Extracted and purified DNA is enzymatically digested to generate a series of fragments that are size-separated by gel electrophoresis, then transferred to a membrane (usually nylon) in denatured form. Specific DNA sequences present in the restricted fragments can be identified by hybridization with a radioactively labeled DNA probe of complementary sequence, by washing off the unbound probe and then visualizing the specific fragments and overlaying x-ray film. This was also known as restriction fragment length polymorphism (RFLP) analysis. The basic forensic analysis can be simply explained in ten steps to generate an autoradiograph (see Figure 8.1).

8.2.2 The First DNA Case

When DNA fingerprinting was being initially published in the scientific literature, a couple of U.K. forensic scientists (Peter Gill and Dave Werrett at the Forensic Science Service's Central Research Establishment in Aldermaston) were working with Dr. Jeffreys to apply DNA technology to stains typically found as forensic biological evidence. They successfully obtained the characteristic multibanded profiles from many possible sources of potentially forensically significant biological evidence, including bloodstains, semen, and vaginal fluid and hair roots. Besides successfully DNA typing single contributor samples, they also devised a strategy to enable spermatozoa to be typed in a mixture with vaginal epithelial cells. This process (called differential extraction) exploits a property of sperm nuclei enabled by the cross-linked thiol rich proteins for cellular protection against sodium dodecyl sulfate/proteinase K treatment that easily lyses the female epithelial cells. After the initial treatment, the still intact sperm can be separated from the female origin DNA by centrifugation and removal of the supernatant. This work was published in *Nature*[7] and became part of the series of Jeffreys' articles in 1985 on applications of this new DNA technology. The differential extraction approach has survived many changes of DNA typing in the crime laboratory and has proven to be a substantial challenge to replace with a technique amenable to automated approaches.

The first DNA case resulted in identifying the culprit who committed two homicides of young girls in a small community of Narborough, Leicestershire, England in 1983 and 1986. Semen samples taken from both bodies had the same ABO and enzyme types (present in about 10% of the population). Consequently, the police suspected the same assailant in both cases. Local police had a suspect who confessed to the second but not the first murder. Investigators decided to approach Alec Jeffreys to see if he could establish the

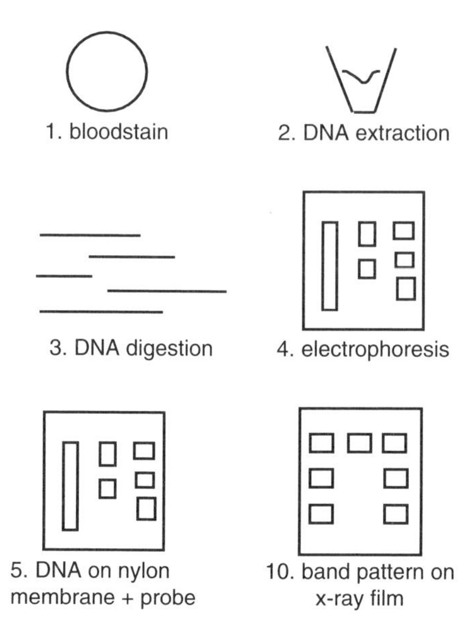

Figure 8.1 Forensic RFLP analysis: Step 1 is the finding and identification of biological evidence that will be a source of DNA. Step 2 is the collection of the material containing the DNA and the purification of the DNA. The amount and condition of the extracted DNA will also be determined. It is important to determine the amount and condition of the extracted DNA before the Southern transfer process and the subsequent hybridization. If the DNA is less than 500 ng or degraded, then the chances for success are diminished. Step 3 is the restriction enzyme digestion of the DNA. The digested fragments are then separated based on size by agarose gel electrophoresis, Step 4. Then a Southern transfer process results in the transferal of size-separated fragments to a nylon membrane in a denatured state (Step 5). The next step is hybridization of a radioactively labeled probe to the membrane (Step 6). A probe that has not hybridized to its complementary sequence (unbound excess) is then washed from the membrane (Step 7). The overlay of x-ray film (Step 8) results in the generation of an autoradiograph image (Step 9). The visualized restriction-digested products or bands of the evidence samples are compared to reference samples (Step 10) to determine a match or nonmatch.

suspect as the culprit in both murder cases. The results of the analysis supported the suspicion that the culprit was the same in both murders and eliminated the initial suspect. These events were of widespread historical importance as a technological advance that affected society by changing the way police could investigate violent crimes. Another consequence of having a genetic means to identify the assailant was that the investigators requested all male residents between the ages of 17 and 34 years old in the region without alibis to provide blood and saliva samples in order to eliminate them as suspects in the murders. About 5000 men were asked to provide samples, and public pressure worked to make it possible. The 10% who could not be eliminated by the blood type and enzyme profile were scrutinized by further DNA analysis. After six months of work, no profiles were found to match the evidentiary samples. A year after this effort, a local worker was overheard by a police informant bragging that he had provided a sample while masquerading as his friend, Collin Pitchfork. The police eventually obtained a sample from Pitchfork and he was found to match the evidence profile. The new suspect was subsequently found guilty of both murders at trial. The story was excitedly covered by the media and even inspired a book by a well-known American writer, Joseph Wambaugh.[9]

8.3 DNA and Crime Labs

8.3.1 The Forensic Science Service and Cellmark in the U.K.

The publications of Dr. Jeffreys' work, the great potential of human genetic identity testing with DNA, and the initial success in the application to a major criminal investigation in the U.K. had many forensic scientists enthusiastic to move into this area of testing. There was however, the issue of technology propriety. The rights to Dr. Jeffreys' technology were restricted to the forensic division of Imperial Chemical Industries (ICI), called Cellmark. While negotiations between the government and Cellmark were on going, Forensic Science Service (FSS) scientists were working on an alternative to the Jeffreys' core sequence probes. This was a DNA probe to the 3′ alpha globin hypervariable region (HVR) obtained from Dr. Weatheralls' Oxford laboratory. Initially reported and characterized by Higgs et al.[10] and Jarman et al.,[11] the probe was being used to identify a family of minisatellites in a similar way to the Jeffreys' probes. A very "hot" probe to the 3′ HVR could be prepared by producing radioactively labeled mRNA. Eventually the proprietary issues with Jeffreys' probes were settled in the U.K. The FSS laboratories and the London Police Met Laboratory and their DNA service continued to expand and constantly evolve.

The first forensic scientific meeting where one of the U.K. scientists directly involved in DNA fingerprinting traveled to North America was in 1987 at an international forensic science conference[12] in Vancouver, Canada. Dr. Nicola Royle from the Jeffreys' lab gave a presentation on how this technology was being applied to forensic analysis. In the U.S. and Canada, there was both optimism and concern about the challenges to be faced. There were issues for developing a standard robust analytical system and for determining the type of validation work needed and the training of crime laboratory personnel. While there was a general consensus about the exciting potential of forensic DNA typing, there were concerns on how to go about establishing and expanding the service. Later that year, Cellmark brought Dr. Jeffreys to

Germantown, MD, for the official opening of their new laboratory. He also accepted an invitation by the Federal Bureau of Investigation (FBI) to lecture at their Quantico Forensic Science Research and Development facility on September 18. The FBI brought in many forensic scientists from around the U.S. to hear the lecture and be updated about the latest developments.

8.3.2 Lifecodes and Cellmark in the U.S.

Work on the application of DNA technology to forensic science in North America was also occurring while Jeffreys' and his colleagues' work out of the U.K. was being published. In 1986 both Cellmark and a U.S. company called Lifecodes began offering DNA typing service in the U.S. A different approach from the European application would be developed in North America.

Lifecodes started out as a company that offered oncogenic and cytogenetic diagnostic services to hospitals. By September 1987, Lifecodes Inc. had moved into forensic DNA analysis and had completed nearly one hundred forensic DNA cases. Their first case evidence analysis results were successfully admitted into court in 1988, *Andrews v. State* in Florida.[13] A series of four single locus-specific RFLP variable number tandem repeat (VNTR) DNA probes, two sex typing probes, and a bacterial DNA probe had been developed for their forensic service. The product was called the DNA-Print™ Forensic System. They offered and marketed a technology transfer program to the publicly funded crime laboratories for their proprietary materials and training. A number of public crime laboratories followed this route. Much later, in 2001, both Cellmark and Lifecodes became one company, called Orchid Cellmark.[14] There are several private companies in the U.S. that now offer high-throughput database sample analysis, plus forensic nuclear and mitochondrial DNA analysis.

8.3.3 The North American Publicly Funded Crime Laboratories

The effort to develop DNA technology for publicly funded laboratories was largely undertaken by the FBI and headed up by Dr. Bruce Budowle. There was also a U.S.–Canadian collaboration between the FBI and the Royal Canadian Mounted Police (RCMP) to help develop and standardize the technology.[15–17] Key to this effort was collaboration set up between the FBI and Dr. Ray White's laboratory in Salt Lake City, UT. Dr. White's lab had developed an extraordinary number of VNTR probes for linkage analysis in mapping genetic disease traits to identify the gene(s) responsible.[18] These DNA probes were made available to the forensic community without proprietary conditions. The pYNH22 or D2S44 DNA probe became the model system, and HaeIII became the restriction enzyme of choice due to its robustness to methylation and the optimal size range of the fragment generated.[15]

A series of probes were selected and a set of analytical conditions was determined in these studies. The Seattle-based Genelex offered another useful DNA probe that was made available to public crime laboratories. The FBI also developed image analysis software for enabling accurate sizing of bands visualized in the autoradiograph. This allowed precision sizing within a gel and between gels to be determined and was used to set a "match" window. A binning system could then be employed to determine allele frequencies for closely sized alleles for population databases.[17] The FBI developed a training program for U.S. crime laboratories to learn RFLP analyses. For many years crime lab personnel from across the country traveled to the FBI Training Academy in Quantico, VA, to undertake an intensive training program. This training would prepare them for implementing the technology in their home laboratories.

8.4 Hybridization Technologies

8.4.1 Single- vs. Multilocus Probes and Effects of Hybridization Stringency

One of the strengths of the initial Jeffreys' technology was also its weakness. The simultaneous detection of multiple loci at once made the chance of a coincidental match very low. Unfortunately, in forensic work there can be more than one contributor of genetic material in an evidentiary sample (e.g., two sources of blood). Unlike the multiplex polymerase chain reaction (PCR) short tandem repeat (STR) typing systems used in forensic DNA testing today, there were no expected, discrete alleles that could be identified and used to interpret the complex profile.

The approach for implementation of DNA analysis in the U.S. taken by both Lifecodes and the public crime laboratories was to bypass the simultaneous multiallelic, barcode-like detection initially used in the U.K. and go with detecting hypervariable DNA sequences from several single sites one at a time. The expectation of only one or two bands from each person being detected for a single hybridization allowed evidence with more than one contributor to be interpreted. After successive hybridization/stripping cycles commonly involving five VNTR probes, a sex typing probe concluded the analysis. Although the single-locus probes were more useful for a wider range of evidence than the multilocus probes, it took a long time before the analysis was complete, sometimes two to three months if some of the samples were weak. The single-locus approach was also eventually developed and employed by Cellmark in the U.K. and the U.S.

One of the key factors to consider in Southern analysis is the stringency of the hybridization and the subsequent wash conditions. A single-locus probe

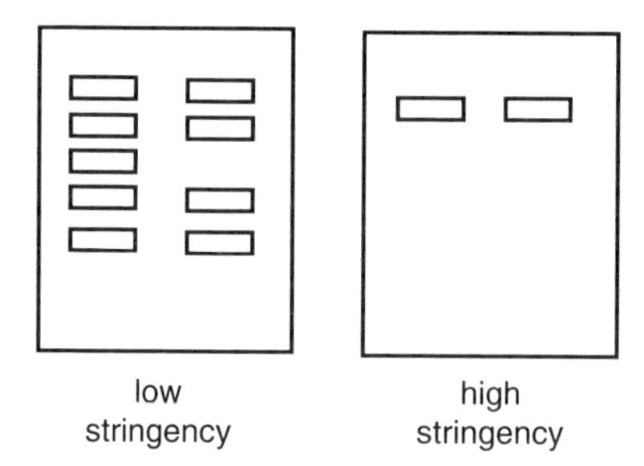

Figure 8.2 The effect of hybridization stringency on RFLP analysis. The results of low-stringency washing and high-stringency washing using 3′ HVR DNA probes on the same Southern transfer membrane are depicted here. The low-stringency wash leaves the probe bound to similar sequences at multiple genome locations, whereas the high-stringency wash leaves the probe only at the best sequence match, a single-locus result.

can be used under low stringency conditions for hybridization and washing to produce a multilocus result. When the blot is washed at high stringency, only one or two bands will show up (see Figure 8.2). The original Jeffreys' probes had other limitations besides in the interpretation of mixed origin biological evidence. There were also some potential problems or at least difficulties in consistently detecting some of the weaker bands of a complex multilocus profile. This was likely due to problems in consistently maintaining the required conditions for the stringency of hybridization and washing.

8.4.2 DNA Quantification

The question of whether general nonspecific DNA quantification or species-specific DNA quantification would be a requirement or an option was an early issue in the development of human forensic DNA typing. The use of single-locus probes enabled the interpretation of biological evidence with more than one contributor. The amount contributed from each donor is one important factor. Another is consistently getting a good strong result (signal) to interpret with ease. The chances for consistent quality results could be improved by considering the amount of human DNA analyzed. The forensic community recognized that the quantification of human DNA prior to analysis was an important step, and it became a standard requirement for laboratory accreditation. The first method used was a slot blot analysis with a set of known calibration standards and a radioactively labeled primate specific DNA probe, D17Z1.[16] Commercial kits were developed later that used chemiluminescent or colorimetric methods for visualizing the results.[19]

Advances in the application of the PCR (described below) increased the sensitivity of successful DNA typing to beyond the limits of the slot blot quantification methods. The development of PCR-derived methods such as

AluQuant™ by Promega Corporation takes advantage of the large copy number and primate specificity of Alu repeat sequences.[20] The luminescent detection of the product is compared to a set of known calibration standards, and a software program provides an estimate of the quantity of human DNA present. The Alu repeat sequence has also been investigated with real-time PCR methods.[21] This application has the additional advantage of providing an indication of how well the sample will amplify. If a Y-chromosome-specific marker is also quantified, then the amount of male DNA present can be determined prior to polymorphic marker analysis.

8.5 The Polymerase Chain Reaction

8.5.1 PCR Origins

In the same year that Alec Jeffreys published his first papers on DNA fingerprinting, the first paper that employed PCR to synthesize copies of DNA sequence was also published.[22] The utilization of a heat-resistant DNA polymerase from a hot springs bacteria (*Thermus aquaticus*) a couple of years later[23] and the development of a specific instrument (thermocycler) to regulate the temperature of the reaction made the technique one of the most significant and widely adapted analytical advances of the time. Kary Mullis was recognized for his role in developing the concept of PCR by being awarded a Noble Prize for chemistry in 1993. Mullis' colleagues at Cetus Corporation in California were responsible for making the concept hugely successful and practical.

The PCR process involves the enzymatic synthesis of multiple copies (millions) of a specific DNA sequence in a cyclical manner.[24] The annealing and extension of two oligonucleotide primers that flank the target sequence result in a replication of the bracketed DNA sequence. The first step is a denaturation of the DNA (generally at about 94°C) followed by the hybridization of the primers (generally at around 64°C) and extension of the primers from the 3′ hydroxyl end at 72°C. This process is repeated around 28 to 30 times, resulting in the duplication of the target sequence over a million times due to logarithmic exponential amplification.

8.5.2 First Forensic Application of PCR Technology

The original PCR researchers at Cetus Corporation recognized the impact that PCR could make on forensic science. They published an application of PCR for forensic analysis in 1986 (Saiki et al.[25]) called reverse dot-blot analysis. This system was used in casework since 1986.[26] The adaptation involved the amplification of the HLA DQ alpha DNA sequences and the detecting of single-base substitutions in the sequence by sequence-specific oligonucleotide

(SSO) probes. The amplified region of genomic DNA produced a labeled PCR product by using biotinylated primers. The array of immobilized sequence SSO probes on a nylon membrane could then be challenged with labeled amplified product and visualized with a subsequent color development step. This system was available in a commercial kit called Amplitype DQA1 ™ (in 1991) that was expanded to include a more allele-specific genetic marker system. This expanded system was collectively called Amplitype PM ™ (in 1993). Although sensitive, the technology had limitations reminiscent of the old serology markers in that they were useful for excluding people from being the biological source of certain body fluid evidence but the number of possible contributors based on coincidentally shared types was high. Also the results were often uninterpretable when there was more than one contributor to the evidence. The minisatellite (core repeats of between 9 and 80 bp) VNTR marker, D1S80, was the first PCR amplified minisatellite employed in forensic analysis.[27] It featured numerous distinct alleles and was small enough to be effectively amplified by a PCR reaction.

8.5.3 Short Tandem Repeats

The next major advance in PCR genetic marker forensic applications occurred in the early 1990s with the use of STR markers. A good reference for background on the forensic application of STRs is the NIST STRbase website.[28] The first paper on STRs was from Dr. Tom Caskey's Baylor laboratory in 1991.[29] These are known as microsatellite markers with 2 to 5 bp repeat sequences and can vary in the number of repeats between individuals. They are widely dispersed in the human genome and are very amenable to PCR amplification (size range of product between 100 to 400 bps). Unlike the RFLP VNTRs, the STR markers can be resolved into bands of distinct alleles by polyacrylamide gel electrophoresis (PAGE). The products can be detected by silver staining or through incorporation of fluorescent dye–labeled PCR primers and a laser-enhanced detection system using a fluorescent detection platform such as the Applied Biosystems (model 377 gel-based or model 310 capillary electrophoresis-based Genetic Analysers) instruments. Incorporating a set of fluorescent dye tags allows for the simultaneous amplification of more than one STR locus at a time and is called multiplex PCR analysis. The forensic adaptation of this process can be explained in a step-wise manner (see Figure 8.3). An excellent reference for this technology and the general application of PCR analysis of human DNA in forensic applications is Dr. John Butler's book.[30]

Scientists led by Dr. Peter Gill at the FSS in England initially developed a first-generation multiplex of four STR markers (comprised of TH01, vWA, FES/FPS, and F13A1) that could be analyzed by the Applied Biosystems 373

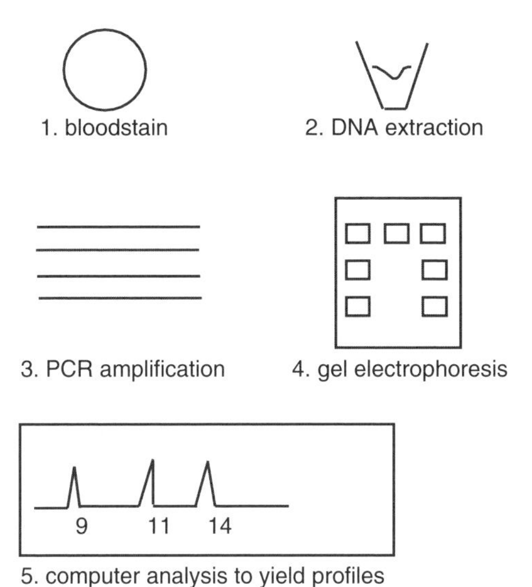

Figure 8.3 Forensic STR analysis: Step 1 is the finding and identification of biological evidence that would be a source of DNA. Step 2 is the collection of the material containing the DNA and the purification of the DNA. The amount of the extracted DNA will also need to be determined since interpretations take into account signal strength. Problems with unbalanced signals can occur from low quantity template (under 250 pg) DNA. Step 3 is the PCR amplification using a multiplex system. Step 4 is capillary (or gel) electrophoresis on an instrument detection platform like the Applied Biosystem Prism® 310 Genetic Analyser (or 377 for gels). Step 5 is the analysis of the run data with specialized software to call the alleles and compare the profiles obtained.

(an earlier model) gel-based electrophoretic detection system.[31] The fluorescent dye-based technology was initially developed for DNA sequencing and the human genome project but has found many applications in the academic world, in clinical laboratories and forensic laboratories. A second-generation multiplex consisting of amelogenin and six STR markers (TH01, vWA, FGA, D8S1179, D18S51, and D21S11) followed.[32] In Canada,[33–38] both the federal RCMP laboratories and Ontario's Centre of Forensic Science Laboratory in Toronto initially employed the FSS quadruplex system (called STR2 by the RCMP). The RCMP added a second multiplex (STR1) consisting of amelogenin for sex determination, FGA, and D21S11. Later, STR1 was modified to include D3S1358 and a third multiplex added (STR3). The additional multiplex was composed of the D5S818, D13S317, and D7S820 loci. These early STR multiplexes did not have allelic ladders, and manual calculations were done to determine if the match criteria had been met. The floating bin approach for the random match probability calculations was employed in a similar manner to the older RFLP method.

In 1996, the FBI, again led by Dr. Bruce Budowle, started an ambitious project involving 22 laboratories to evaluate 17 STR loci and select 13 markers to form the basis for a CODIS core locus set for a future national DNA database.[30,39] The 13 STR markers would provide sufficient discrimination even between close relatives for a national DNA database.[40] At the completion of these concordance studies the 13 loci selected were D3S1358, vWA, FGA, D8S1179, D21S11, D18S51, D5S818, D13S317, D7S820, CSF1PO, TPOX, TH01, and D16S539, plus amelogenin. The introduction of commercial kits with an allelic ladder led more forensic laboratories to switch over to an STR-based analysis. Applied Biosytems out of Foster City, CA, and Promega Corporation, out of Madison, WI, are the primary suppliers to the forensic community.[30] Applied Biosytems also provides the SGM plus kit for the FSS, which consists of ten STR markers plus amelogenin. In 1998, the FBI formally launched the National CODIS DNA database for STR markers. The current kits allow for the option of all 13 STR loci to be amplified in either two separate multiplexes or a single-multiplex amplification.

8.6 Nonhuman DNA Typing

The DNA typing techniques used for human studies apply as well to the study of nonhuman DNA. Advances in the application of PCR technology made it the method of choice for nonhuman DNA applications because it allows for DNA typing from a limited sample. The characterization of nonhuman DNA has been applied to commercial applications (dog, cat, horse, and cattle breeding), enforcement of poaching and wildlife laws (caviar, endangered animals), conservation biology (captive breeding programs for zoos), and patent protection (plant protection for seed companies) as well as for criminal cases (dog,[42–44] cat,[41,44] plant[49]).

8.6.1 Natural Resource Protection

The National Fish and Wildlife Forensic Laboratory in Ashland, OR, has pioneered the use of DNA profiling technology to help with enforcement of federal laws prohibiting the export or imports of illegal animal products.[45] They also provide assistance for wildlife prosecutions for states. In Canada, the province of Alberta has a wildlife protection laboratory that uses STRs in cases of poaching. They provide this service for cases that involve white-tailed deer, mule deer, grizzly bear, and cougar.[46] Forensic applications of STRs to identify illegally caught salmon on the British Columbia coast have been done by a Canadian Fisheries and Oceans department laboratory.[47] Microsatellite genetic markers have been developed to help resolve cases of illegally taken *Sitka* spruce from British Columbia.[48]

8.6.2 Plant DNA Typing

The initial impact of forensic scientific work with plants was their identification by mostly microscopic and or chemical techniques. Similar to the limitations with the serological analysis of human biological evidence, simple identification of plant material may not be sufficient to resolve a criminal matter. Due to the possibility of plant evidence being exposed to harsh environmental conditions or to evidence cover-up efforts by criminals, the established methods of microscopic or chemical identification may be insufficient. DNA analysis could be employed where conventional identification methods do not work well. The possibility of using DNA analysis for identifying illicit plant substances such as marijuana and identification of a specific plant is discussed elsewhere in this book. Besides DNA for routine analysis, there could also be an occasion where an application of DNA analysis could be devised for a specific case.

8.7 Standards and Accreditation

The first working group in the U.S. for public forensic labs to exchange information and discuss setting standards in forensic DNA analysis met in November 1988. The group was funded by the FBI and called the Technical Working Group on DNA Analysis Methods (TWGDAM). Eventually *technical* was replaced by *scientific* and the group became known as SWGDAM.[30] This group has issued several guidelines, and it still continues to aid in setting standards for forensic DNA analysis. There were two National Research Council studies addressing forensic DNA analyses, one published in 1992[50] and the second in 1996.[51] These reports addressed the controversial issues of the time that were frequently argued in court. One of the recommendations was to form the DNA Advisory Board with funding for three years. The DAB group consisted of forensic and academic scientists, and they advised the FBI on the course of action to take for standards and accreditation. They put out guidelines for conducting an audit of a forensic DNA lab. This audit document set the standards for accrediting a forensic DNA lab. SWGDAM and the National DNA Identification System (NDIS) now have assumed the role that the DAB performed. NDIS is responsible for the national DNA databases. The only accrediting body for the U.S. in forensic science is the Association of Crime Laboratory Directors (ASCLD). There are 230 accredited laboratories by ASCLD listed on their website.[52] The only certification of forensic scientists is for criminalists through the American Board of Criminalists (ABC).[53]

8.7.1 Validation Studies for Nonhuman Evidence

A continuous question for forensic laboratories is how to justify allocating personnel time and budgeting of resources for research and validation experiments when pressing caseloads are rarely diminishing. There are two types of validation:

- Developmental validation
- Validation for implementation of existing techniques (internal validation)

A corporation that is preparing for the commercial release of a manufactured product such as a new forensic DNA typing kit typically conducts developmental validation. Validation to implement a preexisting technique at a forensic laboratory means conducting the necessary scientific experiments required for establishing confidence in the performance of a product at that facility.

For nonhuman DNA test methods, very little information and few guidelines have been established or uniformly adopted by the forensic community. In part, this results from the incredible variety of species that exist and may be useful as trace evidence from crime scenes. In reality, the myriad of species and the enormous variety of testing methods have precluded the drafting of standard procedures that will address all situations and concerns when dealing with nonhuman evidentiary samples. In lieu of special guidelines for each species, it is recommended that the general guidelines for human DNA testing be used, where applicable, for validation experiments on plant, animal, and insect species.

The DAB recommends that "a laboratory shall use validated methods and procedures for forensic casework analyses." The DAB further states the following:

- Developmental validation that is conducted shall be appropriately documented.
- Novel forensic methodology shall undergo developmental validation to ensure accuracy, precision, and reproducibility of the test and will include
  a) documentation for the characterization of a genetic locus when possible
  b) tests for sensitivity, species specificity, stability, and ability to interpret mixed samples where appropriate
  c) population distribution data for allelic frequencies and genotypic distribution patterns

- Internal validation shall be performed and documented and will include
  a) reproducibility and precision experiments
  b) the testing of known and nonprobative samples
  c) a qualifying test that will be performed prior to an analyst processing casework samples

d) significant modifications to analytical procedures that will be documented and subjected to validation experiments

- For methods that are not specified by guidelines, the laboratory shall (when possible) select published, scientifically peer-reviewed techniques for application in forensics.

SWGDAM guidelines are tailored to fit specific forms of human DNA testing, including the use of STRs and mitochondrial DNA testing. For specialized forms of testing (e.g., forensic botany), it is recommended that the existing guidelines be adhered to whenever possible. However, if certain types of experiments cannot be performed due to method constraints, it should not be a detriment to acceptance in court.[54]

8.8 Future Trends and Challenges

The advance of forensic DNA technology applications and the growth of the National CODIS DNA databases have increased the demand for human DNA services. This has resulted in large casework backlogs and widespread inadequate resource issues. To increase efficiencies and throughput, the application of robotic workstations for both database samples and casework samples could help to address the backlog problems.[55] One of the most time-consuming tasks of DNA analysis is the evaluation and review of DNA typing data. The development of expert software systems will aid in review and interpretation of data.[56] The tragic events of airplane crashes (e.g., New York and Nova Scotia), acts of terrorism (e.g., World Trade Center disaster), or war crimes (Bosnia) result in a transient great demand for DNA analysis to help identify the victims.[57–60] The technological advances for high-throughput analysis and for working with highly degraded samples helped to meet these challenges. The development of Y chromosome markers, both STRs and single-nucleotide polymorphisms (SNPs), and the successes in mitochondrial DNA sequencing for highly degraded and minimal quantity samples have also helped to increase the chance of obtaining useful information to assist investigators and the courts in resolving criminal matters.[30]

The challenge for DNA technology applications to plants in the forensic arena will be both technical and legal as we have observed for human DNA typing applications. Forensic DNA analysis of marijuana is being developed in the U.S. and elsewhere.[61–64] Establishing casework service in this area could open up other applications for plant DNA analysis. The availability of plant DNA analysis for criminal investigations would provide opportunities to uncover evidence that could resolve difficult cases and be a critical tool in the future.

References

1. Howard, D. and Martin, P., An improved method for ABO and MN grouping of dried bloodstains using cellulose acetate sheets, *J. Forensic Sci. Soc.*, 9, 28, 1969.
2. Saferstein, R., Ed., *Forensic Science Handbook*, vols. 1, 2, and 3, Prentice-Hall, New Jersey, 1982, 1988, 1993.
3. Gaensslen, R.E., *Sourcebook in Forensic Serology*, U.S. Printing Office, Superintendent of Documents, 1983.
4. Sensabaugh, G.F., *Isoenzymes: Current Topics in Biological and Medical Research*, vol. 6, Liss, New York, 1982, 247.
5. Jeffreys, A.J., Wilson, V., and Thien, S.L., Hypervariable "minisatellite" regions in human DNA, *Nature*, 314, 67, 1985.
6. Jeffreys, A.J., Wilson, V., and Thien, S.L., Individual specific "fingerprints" of human DNA, *Nature*, 316, 76, 1985.
7. Gill, P., Jeffreys, A.J., and Werrett, D.J., Forensic application of DNA "fingerprints", *Nature*, 318, 577, 1985.
8. Southern, E.M., Detection of specific sequences among DNA fragments separated by gel electrophoresis, *J. Mol. Biol.*, 98, 503, 1975.
9. Wambaugh, J., *The Blooding*, Perigord Press, New York, 1989.
10. Higgs, D.R. et al., Highly variable regions of DNA flank the human alpha globin genes, *Nucleic Acids Research*, 9, 4213, 1981.
11. Jarman, A.P. et al., Molecular characterization of a hypervariable region downstream of the human alpha globin gene cluster, *Eur. Mol. Biol. Org. J.*, 5, 1857, 1986.
12. Royle, N.J. et al. Presented at the International Association of Forensic Science Conference, Vancouver, Canada, 1987.
13. *Tommy Lee Andrews v State*, 533 So. 2d 851, 1988.
14. Orchid Acquires Lifecodes, press release 2001, see www.ctinnovations.com/news/69.php.
15. Budowle, B. et al., A suitable restriction endonuclease for RFLP analysis of biological evidence samples, *J. Forensic Sci.*, 35, 530, 1990.
16. Waye, J. et al., A simple and sensitive method for quantifying human genomic DNA in forensic specimen extracts, *BioTechniques*, 7, 852, 1989.
17. Budowle, B. et al., Fixed bin analysis for statistical evaluation of continuous distributions of allelic data from VNTR loci for use in forensic comparisons, *Amer. J. of Human Genet.*, 48, 841.
18. Nakamura, Y. et al., Variable number of tandem repeat (VNTR) markers for human gene mapping, *Science*, 235, 1616, 1988.
19. Walsh, P.S., Varlaro, J., and Reynolds, R., A rapid chemiluminescent method for quantification of human DNA, *Nucleic Acids Research*, 20, 5061, 1992.

20. AluQuant™ by the Promega Corporation; see www.promega.com.
21. Nicklas, J.A. and Buel, E., Development of an Alu-based, real-time PCR method for quantitation of human DNA in forensic samples, *J. Forens. Sci.*, 48, 936, 2003.
22. Saiki, R.K. et al., Enzymatic amplification of the β-Globin genomic sequence and restriction site analysis for diagnosis of Sickle Cell Anemia, *Science*, 230, 1350, 1985.
23. Saiki, R.K. et al., Primer-directed enzymatic amplification of DNA with thermostable DNA polymerase, *Science*, 239, 487, 1988.
24. Erlich, H.A., Gelfand, D., and Sninsky, J.J., Recent advances in the polymerase chain reaction, *Science*, 1643, 1991.
25. Saiki, R.K. et al.. Analysis of enzymatically amplified beta-globin and HLA-DQ alpha DNA with allele-specific oligonucleotide probes, *Nature*, 324, 163, 1986.
26. Blake, E. et al.. Polymerase chain reaction (PCR) amplification and human leukocyte antigen (HLA)-Dqa oligonucleotide typing on biological evidence samples: casework experience, *J. Forensic Sci.*, 37, 700, 1993.
27. Kasai, K., Nakamura, Y. and White, R., Amplification of a variable number of tandem repeats (VNTR) locus (pMCT118) by the polymerase chain reaction (PCR) and its application to forensic science, *J. Forensic Sci.*, 35, 1196, 1990.
28. National Institute of Science and Technology (NIST) website for STRs; see www.cstl.nist.gov/biotech/strbase.
29. Edwards, A. et al.. DNA typing and genetic mapping with trimeric and tetrameric tandem repeats, *Amer. J. Human Genet.*, 49, 746, 1991.
30. Butler, J., *Forensic DNA Typing*, Academic Press, San Diego, 2001.
31. Kimpton, C.P. et al.. Evaluation of an automated DNA profiling system employing multiplex amplification of four tetrameric STR loci, *Int. J. Legal Med.*, 106, 302, 1994.
32. Sparkes, R. et al.. The validation of a 7-locus multiplex STR test for use in forensic casework, *Int. J. Legal Med.*, 109, 186, 1996.
33. Fregeau, C.J. and Fourney, R.M., DNA typing with fluorescently tagged short tandem repeats: a sensitive and accurate approach to human identification, *BioTechniques*, 15, 100, 1993.
34. Sweet, D. and Shutler, G., Analysis of salivary evidence from a bite mark on a body submerged in water, *J. Forensic Sci.*, 44, 1069, 1999.
35. Fregeau, C.J., Bowen, K.L., and Fourney, R.M., Validation of highly polymorphic fluorescent multiplex short tandem repeat systems using two generations of DNA sequencers, *J. Forensic Sci.*, 44, 133, 1999.
36. Wickenheiser, R.A., MacMillan, C.E., and Challoner, C.M., Case of identification of a severely burned human remains via paternity testing with PCR DNA typing, *Can. J. Forens. Sci.*, 32, 15, 1999.

37. Richard, M. and Newall, P.J., Validation studies of four short tandem repeat (STR) loci (vWA, THO1, F13A1, FES/FPS) amplified as a multiplex and analysed by automated fluorescence detection, *Can. Soc. Forens. Sci. J.*, 27, 242, 1994.
38. Monaghan, B.M. et al., Implementation of a forensic STR system employing multiplex amplification and automated fluorescence detection, *Can. Soc. Forens. Sci. J.*, 28, 243, 1995.
39. Budowle, B. et al., CODIS and PCR-based short tandem repeat loci: law enforcement tools, *Proceedings of the Second European Symposium on Human Identification*, 73, 1998; see www.promega.com.
40. Hammond, H.A. et al., Evaluation of 13 short tandem repeat loci for use in personal identification applications, *Am. J. Hum. Genet.*, 55, 175, 1994.
41. Menotti-Raymond, M. et al., Genetic individualization of domestic cats using feline STR loci for forensic applications, *J. Forensic Sci.*, 42, 1039, 1997.
42. Shutler, G. et al., Removal of a PCR inhibitor and resolution of DNA STR types in mixed human-canine stains from a 5 year old case, *J. Forensic Sci.*, 44, 623, 1999.
43. O'Neill, M.D., PE AgGen testimony key to Seattle murder convictions; see Biobeat\dog_dna\index.htm, an online magazine at www.appliedbiosys-tems.com.
44. Day, A., Nonhuman DNA testing increases DNA's power to identify and convict criminals, *Silent Witness*, 6, 2001.
45. National Fish and Wildlife Forensic Laboratory in Ashland, OR; see www.lab.fws.gov/.
46. Jobin, R., Patterson, D.K., and Stang, C., Forensic DNA typing in several big game animals in the province of Alberta, *Can. Soc. Forens. Sci. J.*, 36, 56, 2003.
47. Withler, R.E. et al., Forensic identification of Pacific salmonoid tissues to species and population of origin, *Can. Soc. Forens. Sci. J.*, 36, 56, 2003.
48. Newton, C. and Vo, T., Application of microsatellite markers in plant forensics, *Can. Soc. Forens. Sci. J.*, 36, 57, 2003.
49. *State v. Bogan*, 183 Ariz. 506, 905 P.2d 515 (Ariz. App. Div. 1, Apr 11, 1995); see Denver DA's website, www.denverdna.org.
50. Committee on DNA Forensic Science, National Research Council, *DNA Technology in Forensic Science*, National Academy Press, Washington, D.C., 1992.
51. Committee on DNA Forensic Science, National Research Council, *The Evaluation of Forensic DNA Evidence*, National Academy Press, Washington, D.C., 1996.
52. American Society of Crime Laboratory Directors; see website www.ascld.org.
53. American Board of Criminalists; see website www.criminalists.com/abc/A.php.

54. Sensabaugh, G. and Kaye, D.H., Non-human DNA evidence, *Jurimetrics J.* 38, 1, 1998.
55. Mandrekar, P. et al., Automation of DNA isolation, quantitation and PCR setup, *Proceedings of the Thirteenth International Symposium on Human Identification*, 73, 2002; see www.promega.com.
56. Perlin, M.W. and Szabady, B., Linear mixture analysis: a mathematical approach to resolving mixed DNA samples, *J. Forensic Sci.*, 1372, 2001.
57. Leclair, B. et al., STR DNA typing and human identification in mass disasters: extending kinship analysis capabilities, *Proceedings of the Eighteenth International Congress of the International Society for Forensic Haemogenetics*, San Francisco, 1999.
58. Alonso, A. et al., DNA typing from skeletal remains: evaluation of multiplex and megaplex STR systems on DNA isolated from bone and teeth samples, *Proceedings of the Twelfth International Symposium on Human Identification*, 2001; see www.promega.com/geneticproc/usymp12proc/default.htm.
59. Budimlija, Z.M. et al., World Trade Center Human Identification Project: experiences with individual body identification cases, *Croatian Med. J.*, 44, 259, 2003.
60. Ballantyne, J., Mass disaster genetics, *Nature Genetics*, 15, 329, 1997.
61. Miller Coyle, H. et al., A simple DNA extraction method for dried leaf and seed samples of *Cannabis sativa L.*, *J. Forensic Sci.*, 48, 343, 2003.
62. Gilmore, S., Peakall, R., and Robertson, J., Short tandem repeat (STR) DNA markers are hypervariable and informative in *Cannabis sativa*: implications for forensic investigations, *Forensic Sci. Int.*, 131, 65, 2003.
63. Gigliano, G.S., *Cannabis sativa L.* — Botanical problems and molecular approaches in forensic investigations, *Forensic Science Review*, 13, 1, 2001.
64. Gillian, R. et al., Comparison of *Cannabis sativa* by random amplification of polymorphic DNA (DAPD) and HPLC of cannabinoids: a preliminary study, *Science & Justice*, 35, 169, 1995.

DNA and the Identification of Plant Species from Stomach Contents

9

CHENG-LUNG LEE, IAN C. HSU, HEATHER MILLER COYLE, AND HENRY C. LEE

Contents

0-8493-1529-8/05/$0.00+$1.50

9.1 Part I: Ancient DNA

Ancient DNA is defined loosely as DNA obtained from very old samples. It differs in practical terms from "cold case" DNA, which is most often referring to biological samples from unsolved forensic cases from 20 years ago or less. Ancient DNA refers to DNA recovered from archeological remains of bones, insects embedded in amber resin, plant fossils, mummies, and museum skins.[1–3] The molecular biology of DNA from such archived samples has become important in research studies of archeology, conservation biology, and forensic science. In all of these areas of study, the difficulty of obtaining usable and relevant DNA is a common thread that binds them together. Ancient DNA can provide information on the following:

- Development of agriculture from plant remains
- Climatic conditions from plant, animal, and insect remains
- Interpretation of events at individual sites based on identification of kinship and sex typing of human remains
- Broad-scale population migration patterns and colonization
- Last meals eaten in ritualistic deaths

Whenever an event occurs (e.g., catastrophic earthquake, routine burial), pollen and plant fragments are likely to be incorporated into the soil or adhere to the objects at the site. Then, depending on the event and conditions surrounding it, plant fragments or pollen will be maintained as clues to the time of year, the regional plant species and habitat, and local geography.

9.1.1 Forensic Ecology for Dating of Human Remains

There are many different examples where plant fragments and pollen have been useful in adding information to archeological sites and mass graves. A few examples illustrate the interesting uses of plant material in establishing time or season of death. In 1994 in Magdeburg, Germany, a mass grave of 32 male skeletons was discovered.[4] The identity of murderers and victims were both unknowns, as was time of death. Two possible hypotheses were proposed. First, the skeletal remains were from World War II and the Gestapo murdered the men in the spring of 1945. The second hypothesis was that Soviet soldiers were killed by the Soviet secret police following a revolt in the German Democratic Republic in June 1953. Since these two events were known to occur at different seasons of the year, it was expected that pollen content in or among the remains should reflect the season of death and burial. A pollen spectrum was established for each season of the year to provide an independent indicator to compare with the evidentiary samples from the mass grave. The following plant species were found to represent each of the following seasons:

- Spring (alder, hazelnut, willow, and juniper predominate)
- Summer (plantain, goosefoot, sagebrush, and lime predominate)

Some plant species (especially wind-borne pollen) are releasing pollen at different intervals throughout the year. These plants include birch, pine, and hazelnut. These plants are not good seasonal indicators and would therefore not be further analyzed. Twenty-one skulls from the mass grave were examined and the nasal cavities rinsed and the wash solutions collected for pollen analysis. Seven out of 21 skulls had high levels of plantain pollen, an indicator species of a summer season. Therefore, the pollen analysis supported the hypothesis that the human remains were those of Soviet soldiers killed by the Soviet secret police following a rebellion in June 1953.

Natural plant fibers are also informative, especially at archeological sites. FT-Raman microscopy can be applied as a nondestructive type of analysis for identification.[5] A reference library of dyed and undyed plant fibers can be constructed for comparison with samples from burial sites. Some fibers that could be useful include the following:

- Flax
- Jute
- Ramie
- Cotton
- Kapok
- Sisal
- Coconut

Pollen has also been used as supporting evidence for the Shroud of Turin, which is believed to be the burial cloth for Jesus Christ.[6] Max Frei, a Swiss botanist, collected pollen grains from the Shroud in 1973. The plant species that were identified included

- *Gundelia tournefortii* (thistle)
- *Zygophyllumm dumosum* (bean caperl)
- *Cistus creticus* (rock rose)

This particular combination of flowering plant species is most common as native plants in areas of the Middle East, including Jerusalem. These flowers bloom in Israel in March and April. The pollen analysis contradicts a 1989 carbon-dating study that suggested the shroud was 700 years old and originated in Europe.

9.1.2 Dating of Plant Pollen

The oldest verified sample of plant DNA has been isolated from soil deep within the permafrost of Siberia and belongs to grasses, sedges, and shrubs estimated to be between 300,000 and 400,000 years old.[7] Although there is evidence of plants and animals dating back hundreds of millions of years, DNA from such specimens has not been identified because it has degraded into unusable fragments. In a region of northeast Siberia, the research team drilled cores ranging from 2 to 30 m in depth into sediment, between the Kolyma and Lena rivers. Then, they looked for familiar fragments of plant chloroplast DNA. The chloroplast is the energy-generating organelle of a plant cell; it contains its own genetic material that is separate from that of DNA in the nucleus. Ancient pollen is also often found in sediment cores. But pollen can be blown large distances by the wind. By focusing on chloroplasts, the team determined which plants actually grew at the location. They found evidence that grasses, sedges, and shrubs colonized the area until around 10,000 years ago.

9.2 Part II: Types of Plant Evidence Found in the Stomach

Approximately 70 to 100 plant species are eaten in the average human diet. All parts of the plant may be eaten including leaves, petioles, roots, stems, fruits, and seeds. Some commonly ingested plant species are

- Spinach
- Lettuce
- Carrots
- Beets
- Turnips
- Radishes
- Potatoes
- Asparagus
- Apples
- Peaches
- Tomatoes
- Bean species
- Sesame seeds
- Artichokes
- Capers
- Sunflower seeds

Common digestion times include 0 to 2 minutes in the mouth where food is exposed to enzymes such as amylase and pepsin. Food remains in the stomach for approximately 2 to 6 hours in what amounts to an acid bath where mechanical churning and grinding occurs. Food then travels into the small intestine for 2 to 8 hours followed by a 6 to 9 hour stay in the large intestine before it is expelled. The total time for food to be digested is thus 10 to 25 hours.[8] Some factors that affect the emptying of the stomach are the following:

- Caloric content
- Acidity of the food
- Temperature of the food on ingestion
- Electrolyte content
- Level of exercise for the body
- Level of hydration for the body
- Individual genetic variation
- Gender
- Volume of the meal

A decrease in the level of physical exercise will result in a comparable decrease in digestive rates. Females also have reduced digestive rates when compared to most men. In addition, a higher carbohydrate level will slow down digestion.

Traditional methods for the analysis of stomach contents involve the use of a light microscope to observe unknown samples from the evidence and compare to known reference samples in order to make an assessment of total stomach contents. Often, a comparison microscope that allows visualization of two slides (known reference sample and evidentiary sample) simultaneously is helpful. Several steps in rational scientific thinking must occur, including

- Are any of the stomach contents identifiable?
- What are the plant species?
- Are these plant species consistent with statements for the victim's last meal?
- Is the timing or state of digestion consistent with statements?

9.2.1 Evidence Collection

To collect stomach contents, a medical examiner typically scoops out the matter found inside the dissected stomach at autopsy. A standard preservative may be used (e.g., 10% formalin) to denature enzymes and kill bacteria. The use of preservatives is fine for microscopic use, however, for later DNA typing methods may be inhibitory for PCR amplification. It is recommended that a fraction of the stomach contents be frozen for potential DNA usage later in the case. Other types of botanical evidence are also useful to forensic botanists, including

- Plant material found on a suspect's clothing can link him to a field near the house where a murder occurred (path of entry established).
- Examination of a vehicle can identify plant species specific to a certain elevation that can aid in identification of crime scene and location of a body.
- Grass samples can link a suspect to a certain geographic location (e.g., roadside versus river valley).

9.2.2 Collection and Processing of Fecal Matter

Fecal matter may be frozen or stored fresh for a period of a few days at 4°C.[8] To analyze excrement for plant cells and seeds, a dissecting microscope can be used initially. With a sterile forceps, seeds should be removed from other material and rinsed in sterile water. After rinsing, they should be air-dried. To observe plant cells using a light microscope, a few drops of fecal matter should be smeared as a thin suspension on a glass slide. One should allow the slide to dry or gently warm it on a slide heater to "fix" the cells to the slide. The slide should then be placed in 70% ethanol for 5 minutes to begin the dehydration process. Then the slide should be transferred to 95% ethanol for continued dehydration for 5 minutes. Following the same procedure, the slide then needs to be transferred to 100% ethanol for 5 minutes more. As a final step, a histological clearing agent (e.g., HistoClear, National Diagnostics) should be applied for five minutes. The clearing agent will remove any alcohol that may interfere with permanent mounting reagents.

Some cases where fecal matter has been used include robberies and homicides.[9] Fecal matter collected from clothing samples from a homicide suspect were analyzed for plant cell content and compared to the excrement found on the clothing from the rape–homicide victim. Known food reference specimens were used for comparisons. The plant cell composition of the fecal matter on both individuals' clothes showed remarkable similarity. In addition, a similar linkage was made when examining feces-stained clothing from a robbery suspect and comparing to fecal matter left at a crime scene.

9.3 Case Examples[10]

9.3.1 Blueberry Seeds Unravel a Child Abuse Case

In 1997, forensic botanist Anita Cholewa of the University of Minnesota was asked to review a case where a man had been accused of abusing the daughter of his girlfriend. The man told police that he had fed the young girl blueberry pancakes and she had become ill and vomited on him. When the man's blue jeans were examined, blueberry seeds were located on the inside of his pants zipper, which did not support his story. If the child had vomited from illness, the blueberry fragments should have been found only on the outside of his clothing.

9.3.2 Was McDonald's a Last Meal?

In 1982, a young female body was found along the side of an unlit highway. In an attempt to establish a time of death, the stomach contents were analyzed for level of digestion and then sent to forensic botanist Jane Bock at University of Colorado for plant species identification. Botanist Bock teamed up with physiologist David Norris to combine their skills to identify cells from the stomach sample and try and determine a time frame for death. Statements made to police indicated this young deceased woman ate her last meal at the popular food chain McDonald's. Analysis of her stomach contents, however, indicated a meal of cabbage, kidney beans, tomatoes, onions, and peppers. Her death was later linked to a serial killer and her last meal to a restaurant with a salad bar.

9.3.3 Murder of a Husband?

In 1995, Norris and Bock were asked to testify regarding a 1993 homicide of a man suspected of being murdered by his wife in Colorado. The body of Gerry Boggs was found in his home and he was known as the eighth husband of Jill Carroll. Mr. Boggs's stomach contents were analyzed to try and determine the time of death. It was known that the victim always ate eggs, toast, and hash-brown potatoes at the same diner every morning. His stomach contents were indeed consistent with potatoes, even though the medical examiner's report stated them as noodles. The presence of potatoes from hash browns meant that he must have died between 2 and 4 hours after eating breakfast. The wife was present with a new boyfriend later that day but for the time in question had no alibi.

9.3.4 Death by Accidental or Intentional Drowning?

A young mother claimed her infant son was abducted from her while in a public park in California. The mother claimed that she chased the suspect through the park until she could no longer continue running. A few hours later, the baby was found deceased and floating in a small creek. Samples of water were extracted from the infant's lungs as evidence and compared to water from a fountain at the public park and from the creek where the baby was found. From the water from the lungs, freshwater organisms called diatoms were identified. The species identified were consistent with both those found at the park and those from the creek water, indicating that the baby had been submersed in both. After presenting the evidence to the young mother, she broke down and confessed to the accidental drowning of her son in the fountain at the park. Afraid of what had happened, she attempted to make it look like abduction followed by death in the creek.

9.4 Part III: Current Research in Gastric Forensics

9.4.1 Introduction

At the crime scene, sometimes, vomit or feces from the suspect or victim is found. But what types of analyses can be performed for those kinds of evidence? A forensic scientist's most important work is to find a new technique for the identification of evidence and for solving a crime. The individualization of a plant species from stomach contents that came from autopsy, vomit, or feces by using DNA typing methods is a novel area of research. Although many cases have been published on the identification of plant material from the stomach, they have used simple microscopic methods for analysis of cell wall shapes and thickness to classify edible plant species. Many of these cases revolve around supporting or refuting an alibi for events that happened just prior to death in homicides and child abuse situations. Research at the Connecticut State Forensic Science Laboratory is designed to extend the current capability of microscopic plant species identification from stomach contents and feces to individualization of these plants by amplified fragment length polymorphism (AFLP)-DNA typing.

From experiment results, many different kinds of seeds from feces can be recovered. This means that we, as humans, cannot digest those seeds after we eat them. So, we can predict that those kinds of seeds can be collected from vomit or stomach contents too since those are earlier steps in the digestive process. Seeds that have a tough, durable seed coat or membranes are more likely to resist the stomach acid and pass through the digestive system with their DNA intact. Sieve experimental results to recover seeds from feces show seeds that may yield DNA after digestion and include

- Tomato
- Green pepper
- Red pepper
- Watermelon
- Grape
- Rye

As a model system, tomato was selected because it is a popular fruit and it can be eaten raw or boiled. The first step in performing this research was to obtain a sufficient quantity and quality of DNA for further individualization using AFLP. Ten to 20 nanograms are required for AFLP-DNA typing.

Two methods of grinding samples were assessed to determine if DNA yields could be improved. These methods were mechanical grinding by hand versus using an automated Mixer Mill MM300 (QIAgen Inc., Valencia, CA and Retsch Technologies GmbH, Haan, Germany).

9.4.2 Materials and Methods

Prior to DNA extraction, tomato seeds were either ground by hand in liquid nitrogen in a mortar and pestle or pulverized using the Mixer Mill MM300 following the manufacturer's protocol. DNA extractions were performed as recommended by the manufacturer's protocol (DNeasy® Plant Mini and DNeasy Plant Maxi Handbook) for the DNeasy plant extraction kit (QIAGEN Inc.). DNeasy Plant Kits provide a fast and simple way to isolate DNA from seed. The simple DNeasy Spin Column procedure yields pure total DNA for reliable PCR in less than one hour. Purification requires no organic-based extraction or alcohol precipitation, and involves minimal handling. This makes DNeasy Spin Columns highly suited for simultaneous processing of multiple samples. DNA is eluted in low-salt buffer or water, ready for use in downstream applications. DNA purified using DNeasy Spin Columns is up to 40 Kb in size, with fragments of 20 to 25 Kb predominating. DNA of this length denatures completely in PCR and shows the highest amplification efficiency. The DNeasy membrane ensures complete removal of all inhibitors of PCR and other enzymatic reactions. DNA purified with the DNeasy Plant Mini Kit is highly suited for use in demanding downstream applications, including

- PCR
- RAPD analysis
- AFLP analysis
- RFLP analysis
- Blotting methods
- Microsatellite analysis
- Real-time PCR

In the DNeasy Plant procedure, seed material is first mechanically disrupted and then lysed by the addition of lysis buffer and incubation at 65°C. RNase A in the lysis buffer digests the RNA in the sample. After lysis, proteins and polysaccharides are salt-precipitated. Cell debris and precipitates are removed in a single step by a brief spin through the QIAshredder Column, a filtration and homogenization unit. The cleared lysate is transferred to a new tube, and binding buffer and ethanol are added to promote binding of the DNA to the DNeasy membrane. The sample is then applied to a DNeasy Spin Column and spun briefly in a microcentrifuge. DNA binds to the membrane while contaminants such as proteins and two wash steps efficiently remove polysaccharides. Pure DNA is eluted in a small volume of low-salt buffer or water. DNeasy purified DNA has *A*260/*A*280 ratios of 1.7–1.9, and absorbance scans show a symmetric peak at 260 nm confirming high purity.

Some minor modifications to the manufacturer's protocol for the DNeasy plant mini kit were made for isolation of DNA from single seeds. Prior to initiating a DNA extraction, the following steps should be performed:

- Buffers AP1 and AP3/E may form precipitates upon storage. If necessary, warm the buffers to 65°C to redissolve (before adding ethanol to Buffer AP3/E). Do not heat Buffer AP3/E after ethanol has been added.
- Preheat a water bath or heating block to 65°C.
- Preheat Buffer AE to 65°C.
- Prepare ice and liquid nitrogen.

1. Grind seed under liquid nitrogen to a fine powder using a 2-mL collection tube and a micropestle. Keep the micropestle in the tube and put on ice. (Do not allow the sample to thaw.) Continue immediately with step 2.
2. Add 400 μL of Buffer AP1 to clean the micropestle (discard the micropestle), then add 4 μL of RNase A stock solution (100 mg/mL) to the tube and vortex vigorously. (No tissue clumps should be visible. Vortex or pipet to remove any clumps. Clumped tissue will not lyse properly and will therefore result in a lower yield of DNA.)
3. Incubate the mixture for 10 min at 65°C. Mix using the Mixer Mill MM300 at maximum speed for 30 seconds after incubation by inverting tube.
4. Add 130 μL of Buffer AP2 to the lysate, mix, and incubate for 5 min on ice. Centrifuge the lysate for 5 min at 13,200 rpm.
5. Apply the lysate to the QIAshredder Mini Spin Column placed in a 2-mL collection tube and centrifuge for 2 min at 13,200 rpm.
6. Transfer flow-through fraction from step 5 to a new 2-mL collection tube without disturbing the cell-debris pellet. Typically 500 μl of lysate is recovered. In this case determine volume for the next step.
7. Add 1.5 volumes of Buffer AP3/E to the cleared lysate and mix by pipetting. It is important to pipet Buffer AP3/E directly onto the cleared lysate and to mix immediately.
8. Apply 650 μL of the mixture from step 7, including any precipitate, to the DNeasy Mini Spin Column sitting in a 2-mL collection tube. Centrifuge for 1 min at 8000 rpm and discard flow-through. Reuse the collection tube in step 9.
9. Repeat step 8 with remaining sample. Discard flow-through and collection tube.

10. Place DNeasy Mini Spin Column in a new 2-mL collection tube, add 500 μL Buffer AW to the DNeasy Mini Spin Column, and centrifuge for 1 min at 8000 rpm. Discard flow-through and reuse the collection tube in step 11.
11. Add 500 μL Buffer AW to the DNeasy Mini Spin Column and centrifuge for 2 min at 13,200 rpm to dry the membrane. This spin ensures that no residual ethanol will be carried over during elution. Discard flow-through and collection tube.
12. Transfer the DNeasy Mini Spin Column to a 2-mL microcentrifuge tube and pipet 50 μL of preheated (65°C) Buffer AE directly onto the DNeasy membrane. Incubate for 5 min at room temperature (15 to 25°C) and then centrifuge for 1 min at 8000 rpm to elute.
13. DNA yields may be visualized by electrophoresis of an aliquot of the sample on a 1% agarose gel and staining with ethidium bromide.

9.4.3 Results

Tomato seeds, although small, support the feasibility of performing DNA testing on stomach contents. From the experimental results, high-quality DNA could be extracted from only half a seed (a minimal sample input). Due to genetic inheritance patterns, seeds should be dissected prior to DNA typing when possible. The seed coat is typically of maternal genetic origin, while the embryo (F1 generation) represents the new genetic profile due to genetic recombination from ovule and pollen during fertilization. The average yield from a single tomato seed that was dissected into an exterior seed coat (SM) and interior embryo (SW) was 62.5 ng/50 μL for the embryo and 100 ng/50 μL for the seed coat (see Figure 9.1). Red and green pepper seeds

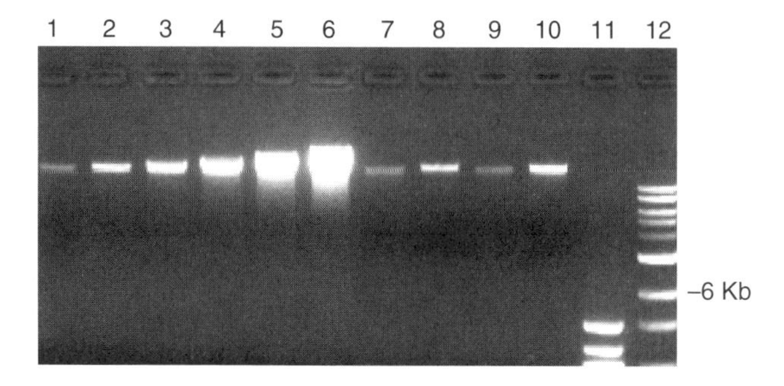

Figure 9.1 The average DNA yield from single tomato seeds after DNA extraction of the embryo with the QIAGEN DNeasy plant extraction kit is depicted on an agarose gel stained with ethidium bromide and visualized under ultraviolet light. Lanes #1–6 are DNA standards of known quantity in ng (12.5, 25, 50, 100, 200, 400); lanes #7–12 are 10% of DNA extract from each seed sample. The average yield is approximately 10 to 12 ng.

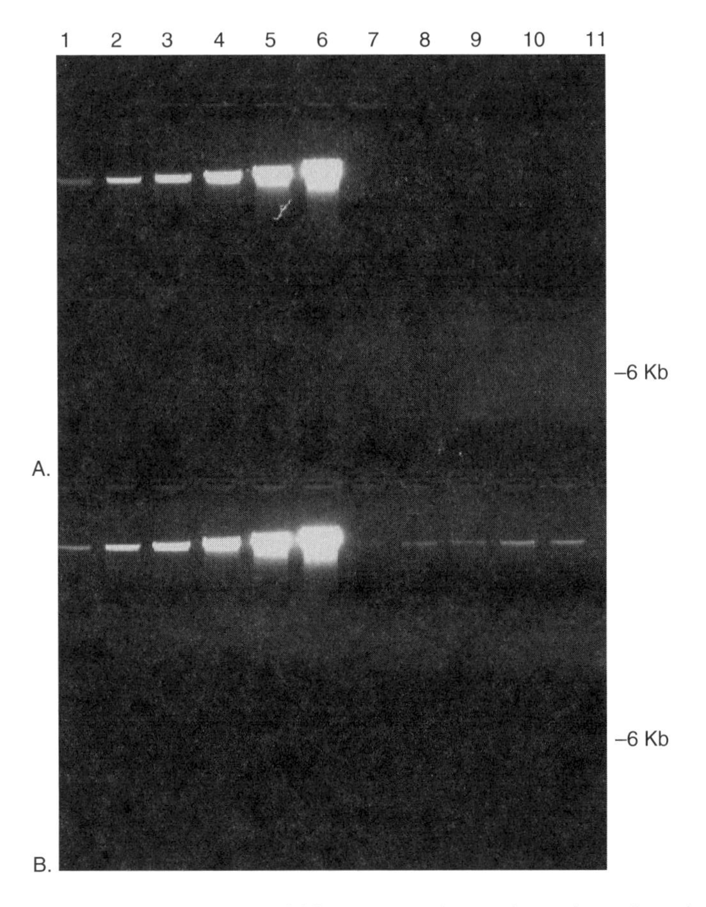

Figure 9.2 The average DNA yield from A) mechanical grinding of single tomato seeds with the QIAGEN Mixer Mill 300 and B) hand grinding of single tomato seeds is shown. In this case, the DNA yield was better with hand grinding because the small seeds would get trapped in the "dead" space between the tube wall and tube cap. With most other tissues, the mechanical grinding step improves DNA yield. Lanes #1–6 are DNA standards of known quantity in ng (12.5, 25, 50, 100, 200, 400); lanes #7–11 are 10% of DNA extract from each seed sample.

have been noted to be similar in seed morphology so similar DNA yields may be obtainable from those seeds. To compare and contrast DNA yields, two methods (manual grinding versus mechanical disruption) were tested for grinding seeds before DNA extraction. The rounded bottom of the extraction tube caps for the Mixer Mill MM300 allows for an air space between the cap and the sidewall of the tube. Due to frequent seed capture in the cap of the tubes for the Mixer Mill MM300, hand grinding yielded better results (see Figure 9.2). This was simply due to inefficient grinding once the seed was captured in the tube cap so the seed could not be reduced to powder by the bead.

In future experiments, the seed DNA extracts will be typed by the AFLP technique to illustrate that PCR amplification is possible. AFLP analysis and the identification of tomato variety-specific markers will allow one to individualize and potentially identify the seed's source by generating a DNA fragment profile. This type of study may help law enforcement and scientists to solve additional cold cases and has application to forensic evidence in crime investigation in the future.

9.5 Summary

The use of forensic botanical evidence from crime scenes for DNA typing is still in the development stages for difficult samples, such as plant matter from stomach contents and excrement. Although microscopic analyses have been performed and have provided investigative information to scientists and law enforcement as illustrated by several case examples in this chapter, DNA typing to individualize samples after species identification will require some optimization of forensic DNA typing procedures. There exists great potential for DNA individualization tests from seed material found in stomach contents, vomit, and feces. Because the seeds are protected from digestive enzymes, they retain usable DNA within the internal embryo tissue. To extract the DNA, the seed is cut open with sterile forceps and a razor blade, and that tissue is further extracted for DNA. Once seeds have been identified that yield sufficient quantity and quality of DNA for further testing, AFLP analysis may be performed. It is envisioned that AFLP or mitochondrial DNA typing can be performed to link a seed back to a parent plant or fruit to aid in police investigations.

As part of the optimization and development of this area of forensic science, a DNA extraction procedure for seeds likely to be recovered from the human digestive tract has been presented. There is a significant window of opportunity for forensic scientists to develop and validate new procedures to utilize this form of evidence as many crime scenes have human excrement associated with them. It is even possible to perform human DNA typing on fecal material to often identify the depositor of the sample. In combination with forensic botanical analyses, it may be possible to provide an investigative lead as to the recent locations that the depositor may have visited through a review of food items and pollen found in excrement left at the crime scene.

References

1. Brown, T.A. and Brown, K.A., Ancient DNA: using molecular biology to explore the past, *Bioessays*, 16, 719, 1994.

2. Burger, J. et al., Paleogenetic analysis of prehistoric artifacts and its significance for anthropology, *Anthropol. Anz.*, 58, 69, 2000.
3. Cooper, A. and Poinar, H.N., Ancient DNA: do it right or not at all, *Science*, 289, 5482, 2000.
4. Szibor, R. et al., Pollen analysis reveals murder season, *Nature*, 395, 449, 1998.
5. Edwards, H., Farwell, D., and Webster D., FT Raman microscopy of untreated natural plant fibres, *Spectrochim. Acta A Mol. Biomol. Spectrosc.*, 13, 2383, 1997.
6. Shroud of Turin, see www.jesusisreal.org.
7. Willerslev, E. et al., Diverse plant and animal genetic records from holocene and pleistocene sediments, *Science*, published online, see www.sciencemag.org/.
8. Workshort (Bock, J., chairperson): an introduction to the applications of forensic botany in criminal investigations. Presented at American Academy of Forensic Sciences, 2001.
9. Norris, D. and Bock, J., Use of fecal material to associate a suspect with a crime scene: report of two cases, *J. Forensic Sci.*, 45, 184, 2000.
10. Mertens, J., Forensics follows foliage, *Law Enforcement Technology*, 62, 2003.

Plant Identification by DNA

Part I: Identification of Plant Species from Particulate Material using Molecular Methods

ROBERT A. BEVER

Contents

There are many types of forensic evidence that may contain trace plant material in the form of leaf fragments and pollen. In addition, evidentiary samples of dust can be informative to reconstruct potential geographic locations for where the evidence may have been transported. As with many fields of forensic testing, timing of deposit is always a concern; however, controlled experiments can sometimes lend information regarding the expected time frame for a pollen deposit. Also, a preponderance of one plant species versus another may give an indication of the season or geographic location where these plants/trees are blooming at a given time. Some types of evidence that

0-8493-1529-8/05/$0.00+$1.50

may contain trace plant material include envelopes, videotapes, clothing, building material, packaging material, shoes, carpets, and interior and exterior areas of automobiles.

The identification of botanical trace evidence has been a valuable tool in forensic science to place individuals at crime scenes and to infer geographic sources of physical evidence. Historically, the identification of botanical trace evidence has relied on morphological and histological analyses. Since the advent of the polymerase chain reaction (PCR) and the compilation of plant DNA sequence information into databases (e.g., GenBank) by molecular taxonomists, DNA-based techniques for the identification of a plant species are possible. The NCBI GenBank database contains thousands of plant DNA sequences that may be useful for species identification. For evolutionary and taxonomy studies, several plant loci have been identified and are routinely used for identification. These loci include the nuclear genes designated 18S, ITS1, and ITS2 (Internal Transcribed Spacer region) and the chloroplast genes coding for ribulose 1,5 bis-phosphate carboxylase (rbcL), ATP synthase Beta (atpB), and NADH dehydrogenase subunit 5 (ndhF). One of the most extensively used loci is the rbcL gene; this locus is approximately 1400 base pairs long, moderately conserved across different species, and has been reported for over 6500 plant species.

10.1 Methods Strategy

The strategy for the molecular identification of botanical trace evidence at the species level is as follows:

- Collection of particulate material from evidence items using a vacuum filtration device or with forceps.
- Extraction of DNA from the particulate material — the particulate material can be a single particle collected with forceps, or a heterogeneous collection of particulates (dust). Methods using bead-beating techniques combined with solid phase extraction techniques have been successful for the extraction of DNA from fresh and aged plant material (e.g., plant material dried for 21 years) and from small amounts of plant material (e.g., a 1.5-mL punch of a dried rose leaf). Our laboratory has had success with both the QIAGEN DNeasy Plant Kit and the MoBio UltraClean Plant Kit. The DNA purified from the particulate material usually represents a mixture of plant material.
- PCR amplification of DNA from particulate material — amplification of the rbcL locus using universal rbcL primer sets or the ITS locus using several different ITS primer sets.

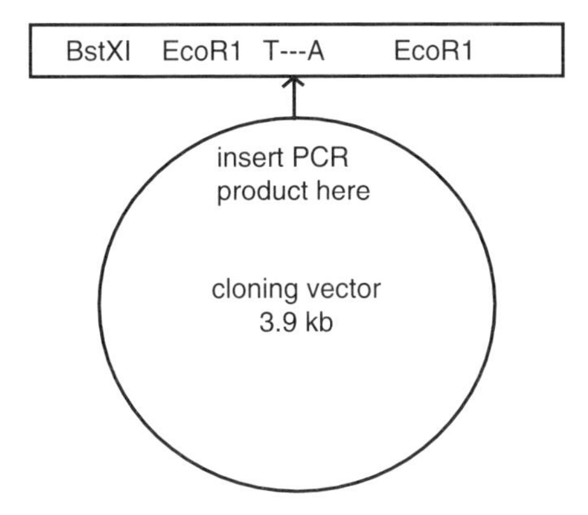

Figure 10.1 The TA cloning strategy for PCR products is schematically illustrated. During the polymerase chain reaction process, an additional adenine is added to the ends of each replicated DNA strand. This allows for the simple and efficient cloning into plasmid vectors that have a complementary thymine for PCR product insertion. The inserted PCR products can then be DNA sequenced to discern the order of the nucleotide bases.

- Use of cloning techniques to separate and identify individual plant components that may contribute to a mixture of plant trace evidence. Cloning of the amplified rbcL product or ITS product is achieved through standard molecular biology techniques and insertion into a cloning vector. The original TA Cloning kit (Invitrogen Corp, Carlsbad, CA) was capable of cloning amplified PCR products from as little as 50 pg of genomic DNA (see Figure 10.1).
- Isolation of individual clones which contain either the rbcL or ITS2 amplified product. The cloning strategy allows the selection of clones with insert (colorless colonies); these clones are screened for presence of an amplified product by restriction enzyme analysis or DNA sequence analysis.
- Sequencing the clones that contain amplified product — DNA sequencing is performed using the rbcL primers along with the Perkin Elmer Dye Terminator Ready Reaction Cycle Sequencing Kit using the ABI 373 sequencing system or the ABI 3100 Genetic Analyzer. Typically two hundred colonies are selected and then sequenced to obtain a variety of DNA sequence data corresponding to the potential mixture of plants associated with the particulate material.
- Use of a computerized BLAST search query to compare the evidentiary unknown sample to known reference samples in the NCBI GenBank database.
- Identification of botanical trace evidence through phylogenetic analysis software to select candidate rbcL sequence matches or ITS sequence matches to a plant species within the NCBI sequence database.

- Interpretation of the plant material associated with the particulate material obtained from the physical evidence. The plant material identified from the physical evidence is compared with the plant material associated with the putative crime scene. Alternatively, the identified plant material can be used as an investigative lead to infer a potential geographical location.

Once an evidentiary sequence is obtained from a sample, identification can be made based on sequences available in the NCBI database. Using the basic local alignment search tool (BLAST), an evidentiary sequence can be compared to thousands of accessioned sequences from previously identified plants. BLAST is a web-based program (http://www.ncbi.nlm.nih.gov/BLAST/) that yields a list of organisms contained in the database with a sequence similar to the sequence of interest; the retrieved data starts with sequences having the greatest similarity. The matches can either be 100% or some lower percentage. Depending on the sequence of interest and the plant group that it belongs to, a high percentage match (near 100%) suggests a possible identification to the species level. However, if the match percentage is less than 100%, phylogenetic analysis is required to better characterize the sequence.

Phylogentic methods are based on evolutionary changes in DNA sequences and are used to determine relationships among different organisms. The aligned sequences are used in parsimony analysis programs such as PAUP,[1] which group taxa based on homologies (characters shared because of a recent common ancestor). Because the NCBI database contains a sample of representative plant species, phylogenetic methods are useful for determining or confirming the genus or family to which a sequence belongs. By constructing a nucleotide alignment that includes the evidentiary sequence and closely related sequences (of known identity), the evidentiary sequence can be classified within a genus and/or family. The greatest amount of information is obtained when a sequence is identified to the species level, but useful data are also obtained from genus or family level identification. The global distribution of most plants is well understood. After a sequence has been identified to species, genus, or family, the distribution data can be used to infer a location or correlate to a putative crime scene. Often, more than one plant is associated with a piece of evidence. The more plants that can be identified from a sample, the more distributional data that can be used to infer a geographic location.

10.2 Mock Case Example

An example of this technique is demonstrated from a mock case involving the collection of particulate material from socks worn in an unknown region of the world. This example has been presented at the Twelfth International Symposium on Human Identification.

10.2.1 Materials and Methods

An article of clothing was exposed to an unknown region of the world and sent to our laboratory. Less than 0.1 grams of particulate dust was collected from the clothing using vacuum filtration. DNA was isolated from the dust using the DNeasy DNA extraction kit (Qiagen Inc., Valencia, CA). The purified DNA was used in a PCR reaction. ITS1 and rbcL sequences were amplified using the following primers, ITS1: ITS5; GGAAGTAAAAG TCGTAACAAGG, GRASS-ITS1; GACTCTCGGCAA CGGATATCT, rbcL: Z1; ATGTCACCACAAACAGAA ACTAAAGCAAGT, 895R: ACCATGATT TTTCTGTCTATCAATAACTGC. The PCR products were cloned using the Original TA Cloning Kit (Invitrogen Corp., Carlsbad, CA) and sequenced with M13 primers following the standard dye-terminator protocol (Applied Biosystems, Foster City, CA). Sequences were placed in BLAST and compared with data in the NCBI database. Phylogenetic analysis as implemented in PAUP was used to further characterize the sequences (see Figure 10.2).

10.2.2 Results

The ITS and rbcL primers successfully recovered numerous sequences from the evidentiary dust (see Table 10.1). However, the primers for each locus did not amplify sequences for the same plants. The chloroplast gene, rbcL, recovered sequences from 3 green algae and 1 angiosperm, while the nuclear gene, ITS, recovered 2 green algae, 9 angiosperms, and 6 fungi, as well as several sequences not similar to anything in the GenBank database.

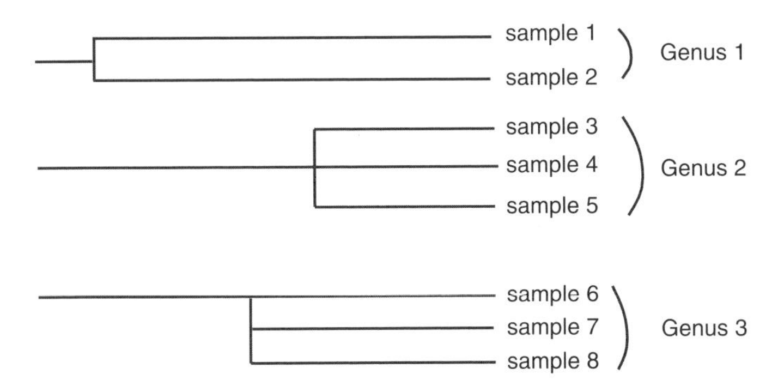

Figure 10.2 Phylogenetic analyses will group sequences of greater similarity or homology together and can be useful for the preliminary identification of a plant species.

Table 10.1 Plant Families and Genera Associated with a Mixture of Particulate Material Collected from a Single Sock

rbcL	ITS1
Green Algae	Green Algae
• *Prasiola* (?)	• Unknown
• *Pleurastrum* (?)	• Unknown
• *Chlorella*	
Poaceae	Asteraceae
• Unknown	• *Artemisia*
	• *Conyza*
	• *Symphyotrichum*
	Brassicaceae
	• *Barbarea vulgaris*
	Fabaceae
	• *Medicago*
	Poaceae
	• Unknown
	• *Triticum*
	• *Poa trivialis*
	Liliaceae
	• *Allium*
	Fungi
	• *Fusarium* (?)

Sequences recovered by rbcL and ITS primers could not all be identified to the species level based on the data available in the current GenBank database. Of the nine angiosperm ITS sequences, two were identified to the species level, five to the genus level, and one was limited to a family-level identification. Four angiosperm families were represented and include Asteraceae (3), Brassicaceae (1), Fabaceae (1), and Poaceae (3). The one rbcL angiosperm sequence could only be identified to the family Poaceae. Green algae sequences recovered by rbcL and ITS could be placed at the ordinal (Volvocales) or class (Trebouxiophyceae) level. ITS also recovered a fungal sequence that is closely related to *Fusarium* and *Gibberella*, two genera known to be crop pathogens. See Table 10.1 for a complete list of recovered taxa.

Of all the organisms recovered by the rbcL and ITS primer sets, only one may be in common with both data sets. Sequences K15 (ITS) and 9B (rbcL) are both closely related to *Pleurastrum* (Trebouxiophyceae).

10.2.3 Discussion

The primary goal of this project was to develop inferences related to the geographic source of trace evidence. In an ideal situation, molecular markers would recover numerous sequences that identify plants with a narrow or endemic range. The data generated from this sample of trace evidence identified several plants (and fungi) with a broad global distribution. However, inferences can still be made as to the environment and ecology common to the organisms recovered. The plants identified are all found in temperate climates and are not likely to be common in dry, alpine, or marine environments. The recovery of *Triticum* and a sequence closely related to fungal crop pathogens suggested an area influenced by agriculture. Identification of *Barbarea* (winter cress), *Artemisia* (ragweed), *Medicago* (medick), and *Conyza* (fleabane) suggested an area recently disturbed, as these plants are early colonizers (weeds) of areas that have rich, damp soil, with direct sunlight. The incidence of green algae in the dust sample is consistent with an environment that is wet or damp and rich in nutrients. Additionally, several rbcL sequences were identified as closely related to *Prasiola* (Chlorophyceae) and may share a similar habitat to this genus; *Prasiola* can be found in moving water, adhering to rocks. The occurrence of the *Prasiola*-like algae suggested the source of the dust might be near the rocky outcroppings of a stream or river. If knowledge exists as to the general origin (region or country) of a sample, as in many cases with forensic evidence, the inferred habitat could be integrated with geography and land use patterns found in that region. In this manner, investigative leads could be generated.

Sequences generated in this study indicate that the rbcL and ITS primers did not recover the same taxa from the dust sample. Only one sequence was noted to have a similar phylogenetic position, as inferred from each gene. The differences in taxa recovered by the two genes may be a result of primer specificity, the copy number of each gene present in the sample, or random error in the cloning process, that is, the cloned PCR products did not reflect the population of sequences in the amplified product. The ability of different primers to amplify the genes of all or most of the plants in a sample is crucial. Valuable plant indicators could go undetected in a sample if the primers do not hybridize to achieve successful amplification; we are therefore pursuing a multiprimer approach. The utility of the ndhF, atpB, and matK genes is currently being investigated, as well as alternative primer sequences for the rbcL and ITS genes.

Regardless of how well the template is amplified in the PCR reaction, the breadth of sequences currently available in GenBank may also be a limiting factor. Of the eight angiosperm sequences recovered from the dust sample, only two were putatively characterized to the species level (see Table 10.1). The remaining sequences were placed to the genus (5) or family (1) level. Our ability to make precise statements of plant distribution and ecology is limited by the level of sequence characterization. To overcome the lack of sequences in the database, the successful interpretation relies on systematists' willingness to share unpublished data and access to plant material stored in herbaria. However, useful information can be inferred from a genus level identification of sequences, especially in genera that contain only a few species.

The need to pursue sequence identification to the species level is largely dependent on the number of sequences contained in the trace evidence and the taxon in question.

10.3 Summary — The Use of Molecular Systematics to Aid in Forensic Investigations

This molecular technique has been used to identify plant material from clothing, videotapes, mailing material, and dust collected from building material. The identified plants have been used to confirm or refute the presumed locations of a variety of evidentiary items and have been used to infer a geographic region of the world.

However, it should be noted that this method is being used for investigative purposes and should be supported or refuted with analyses of additional forensic evidence and techniques. Currently this method is being used to support or refute observations made with microscopy (e.g., The McCrone Research Institute, Chicago, or Stoney Forensic Inc., Clifton, VA). The role of molecular forensic botany is still being defined but is also expanding. The potential exists to complement traditional forensic botanical methods as well as to further the forensic community's ability to identify plant species in particulate matter. This combination of classical and molecular forensic botany provides a powerful tool for generating investigative leads and solving crimes.

Acknowledgement: Technical work was performed by Robyn Wingrove, Emily Hopkins, Lauren Brinkac, and Lisa Barnes of the Bode Technology Group. The authors would like to thank Ed Goldenberg, PhD, of Wayne State University, and Charles Delwiche of the University of Maryland for their scientific discussions.

References

1. Swofford, D.L., *PAUP*: Phylogenetic Analysis Using Parsimony (*and Other Methods)*, Version 4, Sinauer Associates, Sunderland, MA, 2001.
2. Bever, R.A. and Cimino, M., Forensic molecular botany: identification of plants from trace evidence, in *Proceedings from the 12th International Symposium on Human Identification*, 2001. Available at http://promega.com/geneticidproc/ussymp12proc/default.html.
3. Gleason, H.A. and Cronquist, A., *Manual of Vascular Plants of Northeastern United States and Adjacent Canada*, New York Botanical Garden, New York, 1991.
4. Smith, G., *The Fresh-Water Algae of the United States*, McGraw-Hill, London, 1933.
5. Soltis, D.E., Soltis P.S., and Doyle, J.J., *Systematics of Plants II*, Kluwer Academic Publishers, Norwell, MA, 1998.

Part II: Species Identification of Marijuana by DNA Analysis

MARGARET SANGER

Contents

10.4 Molecular Marijuana Identification

Accurate identification of marijuana (*Cannabis sativa*) from evidentiary samples is a necessary tool for drug and law enforcement. Typical chemical and microscopic methods for the identification of *C. sativa* may be inadequate for identifying some small or poor quality samples that may have low tetrahydrocannabinol (THC) levels and/or those that lack defining morphological structures necessary for the microscopic identification of marijuana. A DNA-based assay for the identification of marijuana would provide a useful means of identifying marijuana samples otherwise unsuited to standard methods of identification. This report describes the utilization of DNA analysis for marijuana identification and reports polymerase chain reaction (PCR)-based assays established for a specific nuclear DNA sequence of *Cannabis*. Also, PCR restriction fragment-length polymorphism (PCR-RFLP) assays of chloroplast DNA sequences are described. Each of these DNA-based assays is capable of uniquely identifying *C. sativa*.

To develop a DNA-based marijuana-specific assay for the identification of marijuana, a random amplified polymorphic DNA (RAPD) fragment of approximately 550 base pairs, derived from OP O-13 primed RAPD reactions, was cloned and sequenced (Table 10.2). The RAPD fragment appears to be of nuclear origin, was found common to all tested marijuana samples,

Table 10.2 Primer Pair Sequences Utilized for DNA Amplification

Fragment	Primer 1 (5′-3′, primer name)	Primer 2 (5′-3′, primer name)	Annealing Temp °C	Fragment Size in *C. sativa* (Kb)	Fragment Size in *H. lupulus* (Kb)
GH	gaggtttctaatttgttatgtt (I)	actagaggacttggactatgtc (II)	58	0.20	0.20
CS47-50	aatgggaaaggttactaaaacctttcg (CS47)	attgtatgatctaaacgtcggtac (CS50)	59	0.392	none
CS47-49	aatgggaaaggttactaaaacctttcg (CS47)	caaaagtgtaaacagtgggagagacac (CS49)	62	0.495	0.495
OP 0-13	gtcaaggtcc (0-13)	none	37	0.558	none
HK	acgggaattgaacccgcgca (*trn*H)	ccgactagttccgggttcga (*trn*K)	59	1.75	1.75
CS	ggtcgtgaccaagaaaccac (*psb*C)	ggttcgaatccctctctctc (*trn*S)	58	1.61	1.61
SfM	gagagagagggattcgaacc (*trn*S)	cataaccttgaggtcacggg (*trn*fM)	59	1.25	1.25

and was used to develop a marijuana-specific PCR assay that yields a 392 base-pair product only from PCR amplification of marijuana DNA (fragment CS47-50, Table 10.2). No discernable products were produced from DNA samples from other plant species, including hops (*Humulus lupulus*) and other species closely related to marijuana (Figure 10.3; see color insert following page 238). The specificity of the assay rests on the marijuana-specific sequence of primer CS50.

Assays based on chloroplast DNA sequences have also been explored as an alternative method for identification of marijuana from DNA samples. The consensus-optimized primer pairs of HK, CS, and SfM (Table 10.2) were utilized for the PCR amplification of chloroplast sequences from marijuana, and from a number of other plant species, including close relatives. The resulting PCR amplified fragments were subjected to analysis by endonuclease restriction digestion. Easily discernable restriction patterns unique to marijuana were identified for each of the amplified fragments. A restriction pattern specific to marijuana is illustrated by digestion of fragment HK with *Hin*f I (Figure 10.4; see color insert following page 238). Uniquely identifying restriction patterns were also seen for fragment HK with *Alu* I, for fragment CS utilizing *Alu* I or *Mse* I, and for fragment SfM with either *Hin*f I or *Alu* I.

Identification of marijuana by DNA analysis has the advantage of allowing for identification of samples, which might otherwise remain unidentified by standard chemical and microscopic methods, or of acting as a supplemental analysis to standard methods. DNA analysis allows for the identification of marijuana from all tissue sources, not just buds and flowers, including tissue types not normally amenable to standard identification protocols such as single seeds, leafless stocks, roots, or immature tissues. PCR and RFLP methodologies have been well accepted in court for human DNA analysis, and it is expected that they will prove credible for marijuana. The methods are limited to those samples from which DNA of sufficient quality and quantity can be prepared, and are subject to the same concerns of contamination and controls as for other DNA analyses. In addition, the procedures described earlier are for the purposes of marijuana identification but not the individualization of separate marijuana samples (i.e., they will not create unique DNA patterns when comparing marijuana seizures).

General References

Demesure, B., Sodzi, N., Petit, R.J., A set of universal primers for amplification of polymorphic non-coding regions of mitochondrial and chloroplast DNA in plants, *Mol. Ecol.*, 4, 129, 1995.

Gigliano, S., *Cannabis sativa* L. — botanical problems and molecular approaches in forensic investigations, *Forensic Science Review*, 13, 17, 2001.

Linacre, A.T., Thorpe, J.W., Detection of Cannabis by DNA, International Patent WO 98/24929, 1998a.

Linacre, A., Thorpe, J., Detection and identification of cannabis by DNA, *Forensic Sci. Int.*, 91, 71, 1998.

Part III: Case Study: DNA Profiling to Link Drug Seizures in the United Kingdom

ADRIAN LINACRE, HSING-MEI HSIEH, AND JAMES CHUN-I LEE

Contents

10.5 Case Study

In the year 2001, a warehouse was discovered that contained growing chambers and similar equipment for cultivation of marijuana (*Cannabis sativa*) plants. The U.K. Forensic Science Service (FSS) described the operation as "the most sophisticated ever seen in the United Kingdom." Within a private premise, not far from the warehouse, a large amount of *Cannabis* was also found. South Yorkshire Police seized the samples from the house and also collected specimens from the warehouse. Three men were arrested on conspiracy to propagate *Cannabis* for sale and distribution.

The chemical test performed by the FSS identified the presence of Δ9-tetrahydrocannabinol (THC) in all the plants, both in the warehouse and from the separate seizure in the domestic dwelling. The FSS could not link the haul found in the house to the warehouse operation using traditional methods. There was a need to prove beyond a reasonable doubt that the dwelling contained *Cannabis* grown under the operation of the warehouse.

Figure 10.5 Pictured here is an example of a marijuana plant that might be submitted for case analysis for DNA testing. Entire plants or smaller leaf fragments can be used for testing.

South Yorkshire Police approached the FSS, who suggested that they contact the University of Strathclyde in Glasgow, who were pioneering DNA tests for botanical samples.

In August 2001, nine plants from the warehouse were transported to Strathclyde University along with some plant samples taken from the house. Examples of some of the plants are shown in Figure 10.5. Police and the FSS requested the use of DNA profiling to determine whether a genetic linkage existed between the warehouse plants and the plant samples taken from the house.

10.6 Materials and Methods

DNA was extracted using a chemical (CTAB) method. Quantification of the DNA was performed using agarose gel electrophoresis stained with ethidium bromide. The DNA yield was assessed by comparison to known reference standards. PCR amplification was performed using the specific primers CS1F and CS1R on the flanking sequences that were designed to amplify a single short tandem repeat (STR) locus. The primer sequences for CS1F and CS1R were 5'-AAGCAACTCCAATTCCAGCC-3' and 5'-TAATGATGAGACGAGTGAGAACG-3', respectively. CS1F was labeled with the fluorescent dye chemical name JOE. PCR amplification was performed in a 15-μL volume containing 0.15 μM each of primers, reaction buffer (10 mM Tris-HCl, pH 8.3, 2.5 m*M* MgCl2, 50 m*M* KCl, 0.1% (w/v) gelatin), 250 μM dNTPs, 2.5 units of VioTaq DNA polymerase (Viogene) and genomic DNA. The PCR amplifications were conducted in a 2400 model Perkin Elmer thermal cycler with the following conditions:

- Denaturation at 94°C for 1 min
- Annealing at 55°C for 1 min
- Extension at 72°C for 1 min for 32 cycles
- A 30-min final extension at 72°C

The PCR products were then separated and detected on a 310 Genetic Analyzer (Applied Biosystems, Foster City, CA). The allele values were determined by comparison to an allelic ladder that was generated based on a survey of 108 specimens of *Cannabis sativa.*

10.7 Results

A STR genotype was produced from all of the *Cannabis* plants and the material found in the private house. All the samples shared the same STR type (alleles 14, 17) (Figure 10.6; see color insert following page 238). This was compared to a marijuana database previously generated for this single STR locus. The estimated frequency for observing the allelic values of 14, 17 was calculated to occur at a frequency of approximately 0.06% within the marijuana database.

A report was produced for the court as a statement of witness and used as evidence by the prosecution. This was the first known use of DNA on botanical samples to be introduced into the U.K. courts. In the U.K., evidence is subject to challenge by the solicitor acting for the defendants, who may employ a second scientist to challenge the novel evidence. In this case, the evidence was accepted by the court and was not challenged. A precedent has now been set for the use of DNA profiling on botanical samples in the U.K. The jury accepted that the DNA evidence proved beyond a reasonable doubt that the warehouse was producing *Cannabis sativa* in a large-scale operation, which was then being transferred to the house for distribution. The three men were sentenced at Sheffield Crown Court to more than 28 years in prison for conspiracy to produce *Cannabis.*

10.8 Summary

Although many traditional methods are available for the identification of different plants to the species level, there are special circumstances where the use of leaf morphology, for example, or microscopic features does not yield a confirmatory identification. In these circumstances, some of which are discussed in this chapter, species identification by molecular methods is advantageous. Note that while identification may be made, the probative value will still need to be established on a case-by-case basis. A plant species or mixture of plant species that have a rare geographic distribution pattern or limited range will be of greater probative value than species that are common to many locales. As with many forms of classification for trace evidence, the probative value is related to the relative abundance of an item and the combination of items found as evidence.

The Green Revolution: Botanical Contributions to Forensics and Drug Enforcement*

Part 1: Classic Forensic Botany Cases

HEATHER MILLER COYLE, CARLL LADD, TIMOTHY PALMBACH, AND HENRY C. LEE

Contents

11.1 Introduction

Nonhuman DNA typing methods have been used on occasion for solving criminal and civil casework. Although most forensic scientists are familiar

* Previously published in *Croatian Medical Journal*, 42, 340–345, 2001.

0-8493-1529-8/05/$0.00+$1.50

with methods for human identity testing, the use of plant, animal, and insect evidence is not yet common. This is due in part to a lack of awareness by evidence collection teams who do not necessarily see the value in collecting botanical trace evidence. Prosecutors are also frequently unaware of the potential of botanical evidence to provide linkages between crime scenes and individuals, or they may simply not know whom to contact for assistance. In addition, significant resources are required to construct population databases from the large number of plant species that may be encountered in forensic casework. Forensic botany has a wide range of potential applications, including the identification of vegetable matter in stomach contents to verify an alibi, identification of specific locations of kidnapping victims, and the use of tracking drug distribution networks. The goal of this chapter is to present a review of traditional forensic botany methods (i.e., species identification using morphological characteristics) and casework. In addition, this chapter will discuss some of the new molecular methods (DNA) that are being developed for forensic casework, investigative leads, and drug enforcement.

Forensic botany is defined as the use of plant evidence in court. It is subdivided into several botanical subspecialties, including plant anatomy (the study of cellular features), plant systematics (taxonomy and species identification), palynology (the study of pollen), plant ecology (plant succession patterns), and limnology (the study of freshwater ecology). In the past decade, molecular biology and the use of DNA methods have been important tools to further the research of these disciplines.

11.2 Plant Anatomy and Systematics

Plant systematics is a broad discipline that includes the study of evolutionary relationships between plant species and taxonomy (the identification of plant species). Species identification is a typical first step in analyzing botanical evidence for casework. Plant anatomy uses features such as leaf morphology and tree growth ring patterns to aid in species identification and in performing physical matches of evidence, respectively. The kidnapping and death of Charles Lindbergh's young son in 1932 was the first modern-era case to use such botanical evidence in court. A wooden ladder was used to gain access to the second-story nursery to kidnap Lindbergh's son (see Figure 11.1). Arthur Koehler, a wood identification expert for the Forest Products Laboratory of the U.S. Forest Service in Wisconsin, was able to provide critical evidence against Bruno Richard Hauptman, who was later convicted of the crime.[1] Koehler had an excellent academic record and had provided evidence in several cases prior to the famous Lindbergh trial. His testimony is noteworthy since the use of scientific experts in the mid-1930s was generally

A

B

Figure 11.1 (A) The second story nursery window used in the kidnapping of Charles Lindbergh's son. (B) The hand-constructed wooden ladder used to access the nursery window. Photos are courtesy of the New Jersey State Police.

limited to fingerprints, handwriting, bullet comparisons, and analyses of stomach contents.[1] Koehler first identified the four tree species used to construct the ladder as yellow pine, ponderosa pine, Douglas fir, and birch, via microscopic analysis of wood-grain patterns. Next, Koehler analyzed the tool marks left on the wood from both the commercial planing mill and the hand plane used by Hauptman during the construction of the ladder. Koehler used oblique light in a darkened room to observe the plane patterns left on the wood. Amazingly, he was able to trace the wood by the mill plane marks to a shipment of yellow pine delivered to the National Lumber and Millwork Company in Bronx, New York. The hand-plane marks on the ladder exactly matched those made by a hand plane found in Hauptman's possession. Finally, Koehler compared the annual growth rings and knot patterns on rail 16 of the ladder to a section of wood in Hauptman's attic. The pattern of knots and growth rings on rail 16 exactly matched the exposed end of wood in the attic, supporting the prosecution's position that a section had been removed to construct the ladder. This case exemplifies the use of plant anatomy and plant systematics in providing critical links to Hauptman's involvement in the Lindbergh kidnapping.

11.3 Palynology

Forensic palynology refers to the use of pollen in criminal investigations (see Chapter 14). The major plant groups identified as pollen sources include flowering plants, conifers, and ferns. Ferns technically produce spores instead of pollen but are included in pollen types.[2] Pollen is microscopic and not visually obvious trace evidence during crime scene collection, but is retained on clothing, embedded in carpets, and pervasive in soil. Pollen grain morphology can be used to identify a plant genus and often the species.[3] Crime scenes that are restricted to a few square meters, such as a rape scene or the entry point of a burglary, are good choices for pollen evidence.[2,4] Localized areas have a specific pollen distribution pattern representing the combination of plant species found in the surrounding vegetation. Common pollen types from plants that use wind for distribution (e.g., grass, bracken spores) will be less useful than pollen from uncommon, poorly distributed species (e.g., flax, willow). Insect-distributed pollen is typically deposited within a few feet of the source plant. Pollen analysis consists of species identification and an estimation of the percentage that each plant species represents in an evidentiary sample. A similar pollen composition from shoe-prints and from the shoes that made the prints indicates a strong match correlation.[2] Pollen evidence collected from a burglary entrance and a suspect's shoes, for example, could provide a linkage in a case.

A case that exemplifies the use of pollen in criminal casework is described by Horrocks et al.[5] In Auckland, New Zealand, a prostitute alleged that the defendant had raped her in an alleyway approximately seven meters from his car after failing to pay her in advance for her services. The defendant claimed that he had never been more than one meter away from the car and had not entered the alleyway. Furthermore, he claimed that he had not had sex with the victim and the soil on his clothing was from the driveway area. An examination of the crime scene and the evidence showed no footprints and no seminal fluid stains. A soil sample was collected from the defendant's clothing, the disturbed area of ground in the alleyway, and from the driveway area near the defendant's car. All the soil samples were prepared for pollen analysis by deflocculation with potassium hydroxide, acetylation to remove cellulose and organic matter, and a silicate removal step using hydrofluoric acid. Samples were bleached to remove additional organic matter and analyzed under a microscope for pollen identification and counting. The types of pollens were similar between the two locations, but the amounts of each type were different in each sample. The alleyway contained 76% *Coprosma* (an evergreen shrub) pollen, but the driveway sample contained only 8%. The defendant's clothing contained approximately 80% *Coprosma* and only small amounts of other pollen species. These results support the victim's account of the sexual assault taking place in the alleyway.

Pollen analysis has also been utilized to establish time of death.[6] In Magdeburg, Germany, a mass grave containing 32 male skeletons was discovered in February of 1994. The identities of both the victims and the murderers was unknown. Two hypotheses were proposed: (1) the victims were killed in the spring of 1945 by the Gestapo at the end of World War II, or (2) the victims were Soviet soldiers killed by the secret police after the German Democratic Republic revolt in June of 1953. The ability to differentiate between the spring and summer was critical to solving the case. Pollen analysis was performed on 21 skulls. Seven of the skull nasal cavities contained high amounts of pollen from plantain, lime tree, and rye. All of these plant species release pollen during the months of June and July. Pollen analysis supported the hypothesis that the remains were of Soviet soldiers killed by the Soviet secret police after the June 1953 revolt.

11.4 Plant Ecology

Plant ecology involves studying the growth patterns of vegetation in areas that have been disturbed. These patterns and the vegetative (nonflowering) portion of plants can be useful in estimating time of death.[7] For example, when a body is discovered lying on top of a weed plant with a broken top, useful information can be obtained to define time windows for when the death occurred. A certain amount of shading will eventually kill a plant, so if the weed plant is lacking chlorophyll, a minimum amount of time must have already elapsed. If new shoots are present at the base of the plant, this may establish a second time window. Agricultural research on many plant species has defined the time for new-shoot initiation after the top of a plant has been removed. The length of the new shoot can sometimes establish a third time window. Forensic anthropology sometimes uses plant anatomy to determine an approximate time of death.[8] In one case, the brain cavity of a skull was filled with plant roots. The anatomy and developmental stage of the roots indicated that the plant was approximately one year old, and the plant was putatively identified as *Ranunculus ficaria* L (buttercup family). The predictable stages of plant development were useful in estimating the time that the skeletal remains had been in their present location. The investigators

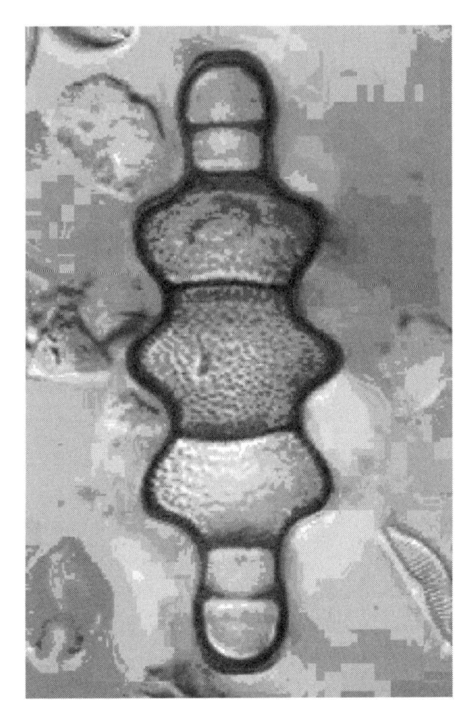

Figure 11.2 Diatoms have a characteristic shape and refractive index that aid in the identification of a species.

were able to determine that the skeleton had been there for at least one year; however, a maximum time could not be established. The plant could have developed secondarily sometime after the body had lain in its present location, so a maximum time estimate was not possible.

11.5 Limnology

Limnology is the study of freshwater ecology and can be applied to a subset of forensic cases. In particular, aquatic plants (e.g., algae, diatoms) have been useful to link suspects to a crime scene or to establish that drowning occurred in freshwater (see Figure 11.2). Diatom populations vary seasonally in lakes, rivers, and ponds.[9] In early spring, diatom populations expand in freshwater. Following this expansion, the live diatoms decline but a large number of dead diatoms remain in summer water. In the fall, a second diatom expansion occurs and then progressively declines through the winter months. When a person drowns in freshwater, diatoms are taken in along with water into the lungs. The diatoms are dispersed to the internal organs of the body. The diatom test is performed by extraction of bone marrow from an intact femur, heating the marrow in a nitric acid solution, and centrifuging to pellet the solids. The solids are examined on slides using phase contrast microscopy for the presence of diatom species. Each species has a characteristic shape and refractive pattern from the silica in the cell wall that can be used for identification. In a study of 771 cases, the diatom test was positive for 28% of presumed freshwater drowning cases but was rarely positive for domestic water drowning. The low rate of diatoms observed in domestic drowning could be traced back to cleaning agents containing crushed diatoms for abrasives.

In 1991, two young boys were brutally attacked by teenage assailants while fishing at a suburban pond in Connecticut.[10] The boys were held at knifepoint, bound with duct tape, and savagely beaten and dragged into the pond to drown. One boy managed to get free, save himself, and rescue his friend. After many hours of criminal investigation, three suspects were apprehended. To link the suspects to the crime scene, investigators seized the sediment-encrusted sneakers of both the victims and the assailants and analyzed them for algal and diatom species. A microscopic analysis of samples from each pair of sneakers plus reference samples from the pond showed the same species and distribution pattern of each species. These results supported the position that the samples all originated from a common freshwater location.

11.6 Plant Molecular Biology and DNA

The previously discussed cases have relied on traditional botanical methods for species identification. In the age of DNA analysis, forensic botany is using molecular biology to aid in criminal and civil investigations. The first criminal case to gain legal acceptance using plant DNA typing was a homicide that occurred in 1992 in Arizona's Maricopa County.[11] A woman's body was found under a paloverde tree in the Arizona desert. Near the body was a beeper eventually traced to a suspect, Mark Bogan. A few seed pods from a paloverde tree were found in the back of Bogan's truck. Officials wanted to know if DNA could match those seed pods to the tree where the body was discovered. Dr. Timothy Helentjaris from the University of Arizona used a technique called randomly amplified polymorphic DNA (RAPD) analysis to generate a band pattern from the evidence in question. He also surveyed a small population of other paloverde trees to determine if the band patterns were unique to each individual. His convincing testimony on plant evidence helped convict Mark Bogan of murder. RAPD marker analysis has also been utilized in civil court cases to identify patent infringements. In Italy, RAPD analysis of a patented strawberry variety "Marmolada" helped settle a lawsuit involving the unauthorized commercialization of the plant.[12]

Molecular methods can be used to identify a plant species from minute leaf fragments and pollen grains. Forensic botanists have utilized DNA technology because often botanical trace evidence does not contain the necessary morphological or histological features that would allow one to identify a plant at the genus or species level. This is particularly true for fragmented and deteriorated plant material. The Bode Technology Group Inc. (Dr. Robert Bever; Springfield, VA) is developing and utilizing molecular methods to analyze botanical trace evidence.[13,14] This type of analysis is a valuable tool for potentially linking an individual to a crime scene or physical evidence to a geographic location. One useful application for the molecular analysis of botanical trace evidence is the identification of a geographic region where a kidnapped individual may be located. Based on flowering times and the plant species represented in the trace pollen evidence found with a ransom note, a geographic region may be identified and would provide the police with an investigative lead. Plant systematists have characterized many loci that are useful for the identification of plants, including several nuclear (18S, ITS1, ITS2) and chloroplast (rbcL, atpB, ndhF) genes.[13,14] Bode Technology Group has identified a DNA extraction, cloning, and sequencing procedure to identify plants using some of those genes. Using these methods, they have identified numerous species of plants from physical evidence. These include species of algae, evergreens, and many flowering herbs, shrubs, and trees. Many plants have a limited geographic distribution or grow in specific habitats. Some of these locations will be general areas, such as roadsides or areas of new construction. Other locations will be more specific, like the Mohave Desert or southern Florida, for plant species that have a severely restricted geographic range. Linking botanical trace evidence to a geographic region could provide law enforcement and investigators with valuable information.

11.7 Drug Enforcement and DNA

Drug enforcement is taking advantage of new plant molecular biology techniques too. Often in drug seizures, identification of the seized substance is a problem, especially if the plant material is fragmented and dried. A variety of methods are currently employed to identify *Cannabis sativa* L (marijuana). Marijuana can be identified by classical botanical characterization, especially using the type of cystolith hairs present on the leaves. However, presence of cystolith hairs is not a conclusive identification since more than 80 plant species have similar cystolith hair morphology.[15] A chemical screening test called the Duquenois–Levine color test[16] is frequently used in combination with cystolith hair observation as a method to identify *Cannabis*. A positive Duquenois–Levine test for marijuana exhibits a purple color in the chloroform layer of the extracted plant material. As with many chemical screening tests, a faint color can be subjectively interpreted. Marijuana can also be identified by chromatographic methods that test for the presence of tetrahydrocannabinol (THC) and other cannabinoids.[17] Unfortunately, not all *Cannabis* samples contain detectable levels of THC. An alternative strategy uses molecular genetics to identify *Cannabis*. *Cannabis* species identification has been achieved by cloning and sequencing the nuclear ribosomal DNA internal transcribed spacer regions (ITS 1 and ITS 2).[18]

In addition to the identification of marijuana samples, it is desirable but difficult to link individual growers and distributors to specific illicit field and greenhouse operations. Molecular genetics may offer a solution to this problem. In certain regions of Canada and the U.S., marijuana is propagated clonally by taking cuttings from a high-THC content "mother" plant and directly rooting them in the soil (Dr. Gary Shutler, RCMP; pers. comm.). This form of propagation results in large numbers of plants having identical DNA analogous to identical twins in humans. DNA typing of marijuana in this situation would allow one to link common growing operations and assess distribution patterns by tracking clonal material. Other growers start their marijuana plants from seed. Each seed has its own unique genetic composition. DNA typing of marijuana grown from seed would allow one to link a leaf found in an individual's vehicle back to a plant from a growing area near the suspect's home, for example.

The Connecticut State Forensic Science Laboratory is developing a molecular strategy for creating unique band patterns from marijuana samples using a technique called amplified fragment length polymorphism (AFLP) analysis. AFLP analysis is based on the selective PCR amplification of restriction fragments from a total digest of plant DNA.[19] The AFLP technique has been used by seed companies to establish patents on novel plant varieties but has not been applied to forensically relevant plant species until recently. Validation of the AFLP technique using marijuana samples and the construction of a marijuana AFLP database for comparative purposes is in progress at the Connecticut State Forensic Science Laboratory.

11.8 Summary

Forensic botany has been useful for criminal and civil cases but is still a much underutilized resource. Forensic scientists are turning to agricultural research to identify molecular methods that can be applied to trace botanical evidence. Validation of those techniques for forensic use is a necessary first step in implementing new molecular tools. In addition to validation, studies surveying the inherent genetic diversity in selected plant populations are desirable.

Plant DNA databases also need to be generated for comparative purposes and to give statistical meaning to a matching profile. Now that human DNA typing methods are commonplace in most forensic laboratories, it is hoped that forensic research will turn to the further development and advancement of nonhuman DNA typing strategies.

References

1. Barni Comparini, I. and Centini, F., Packed column chromatography, high resolution gas chromatography and high pressure liquid chromatography in comparison for the analysis of *Cannabis* constituents, *Forensic Sci. Int.*, 21, 129, 1983.
2. Bever, R. et al., Molecular analysis of botanical trace evidence: development of techniques, American Academy of Forensic Sciences Meeting abstract, Feb. 21–26. Reno, NV, 23, 2000.
3. Bock, J.H. and Norris, D.O., Forensic botany: an under-utilized resource, *J. Forensic Sci.*, 42, 364, 1997.
4. Brinkac, L. et al., Analysis of botanical trace evidence. Proceedings 11th Int. Symposium on Human Identification. Promega Co. Madison, WI Mtg. abstract. 2000.
5. Butler, W., Duquenois-Levine test for marijuana, *J. Assoc. Off. Anal. Chem.*, 45, 597, 1962.
6. Congiu, L. et al., The use of random amplified polymorphic DNA (RAPD) markers to identify strawberry varieties: a forensic application, *Molecular Ecology*, 9, 229, 2000.
7. Graham, S.A., Anatomy of the Lindbergh kidnapping, *J. Forensic Sci.*, 42, 368, 1997.
8. Hall, D.W., *Forensic Botany*, CRC Press, Boca Raton, FL, 353–63, 1997.
9. Horrocks, M., Coulson, S.A. and Walsh, K.A.J., Forensic palynology: variation in the pollen content of soil on shoes and in shoeprints in soil, *J. Forensic Sci.*, 44, 119, 1999.
10. Horrocks, M., Coulson, S.A. and Walsh, K.A.J., Forensic palynology: variation in the pollen content of soil surface samples, *J. Forensic Sci.*, 43, 320, 1998.
11. Horrocks, M. and Walsh, K.A.J., Fine resolution of pollen patterns in limited space: differentiating a crime scene and alibi scene seven meters apart, *J. Forensic Sci.*, 44, 417, 1999.
12. Nakamura, G.R., Forensic aspects of cystolith hairs of *Cannabis* and other plants, *J. Assoc. Off. Anal. Chem.*, 52, 1969.
13. Pollanen, M.S., Cheung, C. and Chiasson, D.A., The diagnostic value of the diatom test for drowning, I. Utility: a retrospective analysis of 771 cases of drowning in Ontario, Canada, *J. Forensic Sci.*, 42, 281, 1997.

14. Quatrehomme, G. et al. Contribution of microscopic plant anatomy to post-mortem bone dating, *J. Forensic Sci.*, 42, 140, 1997.
15. Siniscalco Gigliano, G. and Caputo, P., Ribosomal DNA analysis as a tool for the identification of *Cannabis sativa* L. specimens of forensic interest, *Science & Justice*, 37, 171, 1997.
16. Siver, P.A., Lord, W.D. and McCarthy, D.J., Forensic limnology: the use of freshwater algal community ecology to link suspects to an aquatic crime scene in southern New England, *J. Forensic Sci.*, 39, 847, 1994.
17. Szibor, R. et al., Pollen analysis reveals murder season, *Nature*, 395, 449, 1998.
18. Vos, P. et al., AFLP: a new technique for DNA fingerprinting, *Nucleic Acids Research*, 23, 4407, 1995.
19. Yoon, C.K., Botanical witness for the prosecution, *Science*, 260, 894, 1993.

Part 2: Additional Case Studies

ELAINE PAGLIARO

Contents

11.9 The Hoeplinger Case

On May 7, 1982, the Connecticut State Police Major Crime Squad responded to a call in the affluent town of Easton, CT. Proceeding on foot down the long, crushed-stone driveway investigators noted blood in several locations leading from the drive entrance to the front door. In addition, separate trails of blood drops led to an area of vegetation near the garage and then out of that area to the end of the driveway. Nobody was located outside the house. However, inside the residence of Eileen and John Hoeplinger additional blood patterns pointed to the final resting place of the victim. Eileen Hoeplinger, a young wife and mother of two, was on the sofa in the family room. She was lying face up with much of her clothing in disarray. Bloodstains and blood transfer patterns were noted on pillows and bedding on the couch. Long abrasions were noted on her chin and back. Small amounts of gravel and vegetative matter were also visible on her blouse and skin. The cause of her death was clearly apparent. Eileen Hoeplinger had been struck several times in the head with a blunt object.

John Hoeplinger initially insisted that he had found his wife on the couch. The presence of the abrasions and vegetation on her body and clothing, however, told a different story. On initial examination it appeared that the vegetative fragments were similar to those materials near the blood patterns along the driveway. When this fact was pointed out, the story changed. Hoeplinger stated that he and Eileen had argued the night before and she went to sleep on the couch. When he went downstairs to apologize, she was not on the couch. Hoeplinger claimed that he found his wife outside at the end of the driveway sometime during the early hours of May 7. He then claimed that he carried her back inside and placed her on the couch; some of this movement of Eileen's body must have resulted in the blood drops noted by investigators. He also explained the observed bloodstains on his hands and clothing in this way. There was one problem with his story — the couch where Eileen was found was clearly where she was killed, based on multiple impact blood spatter patterns on the walls and windows near that area. But what weapon was used to kill her and where was it?

Scientists and crime scene personnel noticed an additional trail of blood on the threshold of the sliding doors leading outside from the family room and on the attached deck. More blood drops were on the deck stairs, where a wet T-shirt was hanging on the railing (see Figure 11.3). Several additional bloodstains were noted on the lawn. That blood trail led directly to a small pond behind the house. When the scientists approached the pond (see Figure 11.4), more bloodstains were noted on the rocks at the edge of the water. Near these rocks was a brick, submerged in the water. That brick was similar in construction and color to bricks piled in front of the house. Subsequent laboratory analysis revealed the presence of hairs and tissue embedded in the brick — hairs and tissue that were consistent with those of Eileen Hoeplinger.

But how could scientists link John Hoeplinger with the brick and blood evidence? He continued to insist that an intruder must have come into his home and committed this terrible crime. The answer lay in the T-shirt hanging on the railing. John Hoeplinger had already identified that T-shirt as his garment. He explained that he had rinsed it out after getting blood on it while moving his wife's body back into the house. Scientists at the Connecticut Forensic Science Laboratory carefully examined the T-shirt and noticed several greenish-colored areas. Microscopical examination of those areas revealed deposits of green algae among the diluted bloodstains on the shirt. Those deposits could only have been made if Hoeplinger rinsed out the T-shirt in water that contained algae. What was the logical source of such water? The pond! Samples of pond water were collected and examined at the laboratory. Both forensic scientists and botanists agreed that the algal species found on the T-shirt were the same as those in the pond water. These microscopical organisms provided the linkage between John Hoeplinger and the murder weapon. This microscopical identification and comparison of algae

Figure 11.3 The stained, wet T-shirt hanging over the railing at the Hoeplinger residence on the day of the murder

Figure 11.4 The pond where the murder weapon was found also contained the same species of algae that were identified on the suspect's wet T-shirt.

became a vital piece of evidence in John Hoeplinger's trial. He was subsequently convicted of the murder of his wife.

11.10 Trace Botanical Evidence to Link a Body to a Suspect

On July 7, 1960, a young boy was abducted on his way to school in Sydney, Australia. His parents, Basil and Freda Thorne, received a telephone call demanding 25,000 pounds for his safe return. On August 16, 1960, the young boy's body was located approximately 10 miles from his home. He had been beaten and asphyxiated. His body was wrapped in a rug and mold had begun to grow on his socks and shoes. His clothing had a pink, crusty substance on it as well as some leaves, seeds, and twigs. As the trace evidence was examined, it became clear that hairs obtained from the rug could have come from at least three people and a dog determined to be a Pekingese breed. Botanist Neville White determined that the mold was composed of four types of fungi and had begun to grow on the socks and shoes shortly after the boy's abduction based on the growth and life stages of the fungi. The pink, crusty substance was from a type of mortar used in house facings. One of the seeds was identified as a rare cypress species that was not found in the location where the young boy's body was dumped. After a police broadcast, a postal worker identified a house with a pink color and a cypress tree but the murderer had left with his family on a ship traveling to Sri Lanka. His neighbors confirmed that he had owned a Pekingese dog. As the ship docked, police arrested Istvan Baranyay for murder. On March 29, 1961, he was found guilty and sentenced to life in prison.

11.11 Forensic Archeology and Burial Sites

Forensic archeology often relies on plant matter found in or around burial sites to give information regarding a historically important time period, an indication of important aspects of a society and features of burial rites and body preservation techniques. For example, between 1924 and 1991 (marking the end of communism in the Soviet Union) millions of visitors paid their respects to the embalmed body of Lenin. Two months after Lenin died, two professors (Ilya Zbarsky and Vladimir Vorobiov) mummified his body to maintain him in perpetuity as the great Soviet founder. Mummification was achieved through soaking the body in baths of balsam, glycerine, and potassium acetate. Today, his remains are still remarkably preserved.

Yet another interesting collection of mummified remains can be found in Chinese Turkestan. Some of the mummified remains date back almost four thousand years, which makes them roughly equivalent in age to the better-known Egyptian mummies. What is interesting about these remains of prehistoric people is that they were not Asian, as expected by the geographic region of their burial, but Caucasian in phenotype. These people were most likely nomadic and were tall and blond, with large facial features with round, possibly blue, eyes. There were few gifts of tools or pottery found with the mummified remains, which made it difficult to link these people to any of the known existing cultures of the time; however, their woolen clothing was very well preserved. Most clothing of this type does not survive more than a few centuries, but the mummies of Urumchi have clothing that is brightly colored and woven in intricate patterns that can be used to piece together parts of their past history.

While pond water, balsam and other coniferous resins, and animal- or plant-based textiles are not necessarily the first type of evidence that may be considered in a forensic case, often these key pieces of evidence play a crucial role in adding weight to a circumstantial case or historical mystery.

General References

Barber, E.W., *The Mummies of Urumchi*, W.W. Norton & Co., New York, 1999.

Innes, B., *Bodies of Evidence*, The Reader's Digest Association Inc., Pleasantville, NY, 2000.

Zbarsky, I. and Hutchinson, S., *Lenin's Embalmers*, Harvill Press, London, 1998.

12 Tracking Clonally Propagated Marijuana Using Amplified Fragment Length Polymorphism (AFLP) Analysis*

HEATHER MILLER COYLE, TIMOTHY PALMBACH, CARLL LADD, AND HENRY C. LEE

Contents

* Reprinted in part with permission from Promega Corporation from Proceedings of the Thirteenth International Symposium on Human Identification, Phoenix, AZ, Oct. 7–13, 2002.

0-8493-1529-8/05/$0.00+$1.50

12.1 Introduction

Cannabis sativa (marijuana) is one of the oldest cultivated crops in the world. As a plant, it is valued for both its hallucinogenic and medicinal properties. Marijuana has been used to treat a variety of ailments, including pain, glaucoma, nausea, asthma, depression, insomnia, and neuralgia.[1] Like many cultivated crops (e.g., wheat, corn), it has been domesticated from a naturally occurring weed species and propagated to yield increasing amounts of Δ-9-tetrahydrocannabinol (THC) and for other desirable smoking traits (e.g., flavor, smoothness). Marijuana is a large cash crop in the U.S., and marijuana cases account for enormous asset forfeitures. These forfeited assets, in turn, are used to fund law enforcement budgets for drug-eradication efforts, community-based treatment, and drug-prevention programs. Marijuana is a primary focus for drug-intervention programs in schools and is considered by many as a "gateway" drug to introduce young users to other addictive illicit drugs, such as heroin and cocaine. Most drug arrests are for crimes involving marijuana. In 1998, some 88% of 700,000 drug-related cases were for marijuana possession in the U.S.[1]

Until the mid-1970s, most U.S. marijuana was imported from Mexico. When Mexico joined the U.S. in its marijuana-eradication efforts, a domestic "home-grow" industry was born. Industrious breeders developed several superior *Cannabis* cultivars (e.g., Northern Lights, Skunk #1, California Orange, and Big Bud) in the early 1980s.[1] Most subsequent marijuana cultivars are thought to be derived from these few genetic lines. Constant breeding is in progress, but since it is performed as an underground activity, it is difficult to determine the number of true marijuana cultivars in existence. Numerous seed catalogs are available via Internet access; however, none list genetic markers for cultivar identification. The online seed catalogs rely on traditional morphological descriptors (e.g., red stems, foliage color) complete with photographic examples. The morphological phenotype of *Cannabis sativa* is highly influenced by environmental conditions; therefore, traits such as leaf size and shape are not truly reliable for cultivar identification.

Cannabis sativa is a dioecious, herbaceous annual plant with a 4- to 6-month growing season.[2] Dioecy, by definition, means that pistillate (female) and staminate (male) flowers are presented on separate plants. Marijuana can be propagated in two ways: by seed or by cloning.[2,3] Seeds are a result of sexual reproduction between a pistillate and staminate plant, and produce new individuals with recombinant genotypes. The highly prized Sinsemilla marijuana results from growing pistillate plants in isolation from any source of pollen. Isolated female plants will produce prolific floral buds with a high THC content. Cloning is a form of asexual reproduction that allows for preservation of the genotype due to lack of meiotic recombination. This form of propagation is desirable to the grower because it perpetuates the unique characteristics of the parent plant. It also generates a population of nearly identical, all-pistillate, fast-growing, and evenly maturing *Cannabis* plants. To propagate marijuana by cloning, a cutting is removed from the parent plant and induced to form a new root system.[2,3] Root systems typically develop in 3 to 6 weeks, and the clones are then ready to be transplanted into larger containers. Supplying excess nutrients, carbon dioxide, and light can accelerate plant development. With a sudden shift from 24-h daylight to a 12-h light regime to mimic autumn conditions, marijuana plants can be forced to flower before they are eight weeks of age.[2]

Amplified fragment length polymorphism (AFLP) analysis is a method that can be used for individualizing any single-source biological sample based on a DNA profile.[4] This method has been routinely used for creating highly saturated genetic marker maps for identifying linked traits in plants,[5–14] animals,[15,16] insects,[17,18] fungi,[19] and bacteria.[20–23] Only recently has AFLP analysis been applied to a species in a forensic context.[24–25] AFLP analysis involves the polymerase chain reaction (PCR) amplification of restriction fragments to generate a band pattern that can be used as an identifying profile for the sample. The band pattern can be "adjusted" for complexity based on the use of selective PCR primer sets in the second round of PCR amplification. This method is somewhat analogous to the original forensic restriction fragment length polymorphism (RFLP) DNA typing method that uses multilocus probes for human identity testing. Both AFLP analysis and multilocus probes generate a highly discriminating and complex band pattern for individualizing a biological sample.[26] Scoring complex band patterns and converting them into useful data points is a challenge for the AFLP user; however, the authors have developed a system for categorizing variable peaks that simplifies AFLP analysis.

Forensic analyses involve two steps: identification and individualization of a biological sample:

Identification — For botanical evidence, the molecular identification of a species can be determined by cloning and DNA sequencing of the internal transcribed spacer (ITS) region of the nuclear genome and the large subunit of ribulose 1,5-bisphosphate carboxylase (rbcL).[27] For identification of marijuana, many tests are available, including the microscopical examination of leaf hair morphology, the Duquenois–Levine test, and gas chromatography analyses for THC and other cannabinoid compounds.[28-31] Identification of *Cannabis* by molecular methods has also been published.[32,33]

Individualization — Amplified fragment length polymorphism analysis is a method for individualizing marijuana samples, and the authors are validating this method for profiling marijuana seizure samples. An AFLP database containing profiles from seizure samples obtained from the U.S. and Canada is under construction to establish the extent of genetic diversity in marijuana cultivars. AFLP analysis of marijuana seizure samples can determine if seizures from two or more geographic locations share a common profile. Our data show that AFLP profiles can provide genetic evidence for the form of plant propagation being utilized by the growers of this illicit crop. This type of evidence can be useful in providing investigative leads and for illustrating the extent of grower organizations and distribution networks.

12.2 Material and Methods

12.2.1 Tissue Source

All samples for DNA extraction were obtained from marijuana seizures in the New England area. Fresh (100 mg) or dry (20 mg) young leaf tissue was preferred for DNA extractions; however, stems and flowers were also used. When fertilized female flowers were used, any visible seeds (mature or developing) were removed. If only seeds were available, the perianth and pericarp were removed and the embryo (2 to 5 mg) was processed.

12.2.2 DNA Extraction and Estimate of DNA Yield

DNA was extracted with the QIAGEN (Valencia, CA) Plant DNeasy kit according to the manufacturer's protocol. DNA was eluted from the filter twice with AE Buffer (100 μL) and combined into one tube. DNA yields were estimated by comparing to known mass standards co-electrophoresed with the samples on either a 1.0% or 1.5% agarose gel in 1X TAE buffer. Gels were stained with 0.4 ng/mL ethidium bromide to visualize the DNA. If the DNA was degraded but a high molecular weight (MW) band was visible, that band was used to estimate the quantity. If no high MW band was visible, the DNA was used if the following criteria were satisfied: 1) the majority of the fragments were >3 Kb, and 2) the smear was clearly visible under ultraviolet (UV) light (302 nm) when the extract (2 μL) was loaded into a 4.5 × 1 × 5 mm well (gel dimensions, 14.5 cm length × 11 cm width × 0.75 cm height) and subjected to electrophoresis (100 V, 1 h). If high MW DNA was less than 4 ng/μL, or if degraded DNA was not visible, it was concentrated up to tenfold using a Microcon YM-100 centrifugal filter (Millipore; Boston) and requantified.

12.2.3 AFLP Analysis

AFLP analysis was performed according to the AFLP™ Plant Mapping Kit (Applied Biosystems; Foster City, CA) manufacturer's protocol with the following modifications. Digestion-ligation reactions were prepared in batches of 20 reactions at a time. High MW DNA (approximately 20 ng) was prepared to a final volume of 5 μL in distilled water in a thin-walled 0.2 mL PCR tube and set aside while the other reagents were prepared. For degraded DNA, 5 μL of the DNA extraction was used. The adaptor pairs were annealed by heating in a 90–99°C water bath for 5 min, and cooled at room temperature for 10 min. The enzymes used included MseI (New England Biolabs (NEB) Inc., Beverly, MA: 50 units/μL or Life Technologies Gibco BRL, Rockville, MD: 5 units/μL), EcoRI (Gibco BRL: 50 units/μL) and T4 DNA ligase (Gibco BRL: 5 Weiss units/μL or NEB: 30 Weiss units/μL). All enzymes were maintained in a –20°C cold block while the reactions were being prepared.

In order to prevent pipetting of small volumes of reagents into individual tubes, two "pre-master" mixes were assembled (designated enzyme mix and adaptor mix) for the digestion-ligation step. The enzyme mix was prepared by combining 20 × [1 unit MseI, 5 units EcoRI, and 1 Weiss unit ligase] in the bottom of a 0.5-mL tube and placing it in a cold block. If the enzyme mix recipe required pipetting volumes less than 0.5 μL, then the enzymes were diluted with an appropriate Diluent Buffer (NEB) and the recipe adjusted accordingly. The adaptor mix was prepared by combining 20 × [2.2 μL 5X (or 1.1 μL 10X) Ligase Buffer), 1.1 μL 0.5 *M* NaCl, 0.11 μL 5 mg/mL bovine serum albumin, 1 μL MseI adaptor, and 1 μL EcoRI adaptor]. It was then brought to a final volume equal to (20 × 6 μL minus the volume of the enzyme mix) with distilled water. The following two steps were carried out expeditiously to prevent the adaptors from ligating to one another. The entire adaptor mix was transferred to the tube containing the enzyme mix, and the solution was pipetted gently to mix. The enzyme/adaptor mix (6 μL) was added to each tube containing DNA and gently mixed. The 0.2-mL tubes were incubated at 37C in a thermal cycler with a heated lid for 2 h.

Pre-amplification reactions were performed in batches of 20 to 50 reactions plus one negative control. After diluting the digestion-ligation reactions with 189 μL $TE_{0.1}$ Buffer (20 m*M* Tris-HCl, 0.1 m*M* EDTA, pH 8), an aliquot of each reaction (4 μL) was transferred to a 0.2-mL PCR tube and set aside. The negative control tube was prepared with 4 μL $TE_{0.1}$ Buffer in lieu of the diluted digestion-ligation component. A master mix (designated pre-amp mix) was prepared by combining [the number of samples + 2] × [1 μL AFLP EcoRI and MseI pre-amplification primers and 15 μL AFLP Core Mix]. The pre-amp mix (16 μL) was transferred to each of the 0.2-mL tubes and pipetted gently to mix. PCR was carried out in a Perkin Elmer 9700 model thermal cycler using the temperature profile described in the AFLP kit protocol (ramp speed set to MAX). Pre-amplification reactions were then transferred to 0.5-mL tubes and diluted with 380 μL $TE_{0.1}$ Buffer.

Selective amplifications were carried out in batches of 20 to 50 for each selective primer pair. AFLP profiles were generated with four selective PCR primer pairs (our designation: A1, A4, F4 and F5, where A is EcoRI-ACT FAM, F is EcoRI-AAG JOE, 1 is MseI-CAA, 4 is MseI-CAT, and 5 is MseI-CTA). The diluted pre-amplification reaction (3 μL) was transferred to 0.2-mL PCR tubes and set aside. A negative control was included with each primer pair, using $TE_{0.1}$ Buffer in lieu of diluted pre-amplification components. A master mix (designated selective-amp mix) was prepared by combining [the number of samples + 2] × [1 μL AFLP EcoRI primer, 1 μL AFLP MseI primer and 15 μL AFLP Core Mix]. Expeditiously, the selective-amp mix (17 μL) was transferred to each of the 0.2-mL tubes and gently mixed. PCR was performed as described in the kit protocol.

Amplification products were separated and detected using an ABI Prism 377 DNA Sequencer. A 0.2-mm acrylamide gel (Long Ranger® Singel® Packs, BioWhittaker Molecular Applications, Rockland, ME), a 34-well square-tooth comb, and plates with a 36-cm well-to-read distance were used to separate PCR products. A master mix (designated FLS-ROX) was prepared by mixing [the number of samples] × [4.75 μL Loading Solution (deionized formamide:blue dextran, 5:1) and 1 μL GeneScan-500 size marker [ROX] (ABI)]. FLS-ROX (5 μL) was added to the selective amplification product (4 μL, generated from one primer pair). These samples were heated at 95°C for 2 min, quick-chilled on cold blocks, and loaded (2 μL) into each well of the gel. The gel was run for 5 h at 1680 V.

12.2.4 Data Management

AFLP profiles are represented as peaks labeled with the specified fluorescent dye designated by the selective PCR primer pair. Peaks between 50 and 500 bases were sized with GeneScan® 3.1.2 software. The presence or absence of variable peaks in user-designated categories was determined by alignment of 50 to 100 AFLP profiles and the data converted to binary code with Genotyper® 2.5 software. A binary code database representing each AFLP profile was established and the database was managed using Microsoft Excel software.

12.3 Results

12.3.1 Tissue Source and Presence of Plant Resin Do Not Affect the AFLP Profile

A single hydroponically grown marijuana plant was sampled. The tissues sampled (103 mg, fresh weight) included a female floral bud with resin, a stem (approximately π [1/4] inch thick, no visible resin), and a mature leaf (dark green, no visible resin). The different tissues collected from the individual marijuana plant were found to yield sufficient amounts of genomic DNA for AFLP analysis. Each tissue source exhibited identical AFLP profiles, thereby illustrating that AFLP typing will yield the same profile regardless of the portion of the plant from which the DNA is extracted (see Figure 12.1). The presence of visible resin on the floral bud did not interfere with PCR amplification, as evidenced by the ability to obtain an AFLP profile match with both stem and leaf tissues.

12.3.2 Genetic Evidence of Clonal Marijuana Propagation

Two marijuana-grow operations were suspected to be linked based on informant information. Location #1 had plants of various sizes growing under lights in different containers. Location #2 was a high-tech hydroponic grow operation. Plants from each location were seized and the grower's labeling compared. AFLP analysis of a subset of the seized plants from each location showed that plants with similar labels had identical AFLP profiles for all four selective primer pairs. The AFLP results provide genetic evidence for cloning and demonstrate that the two growers are sharing plant material or are obtaining it from a common source.

Figure 12.1 A) An example of marijuana leaf tissue used for AFLP processing.

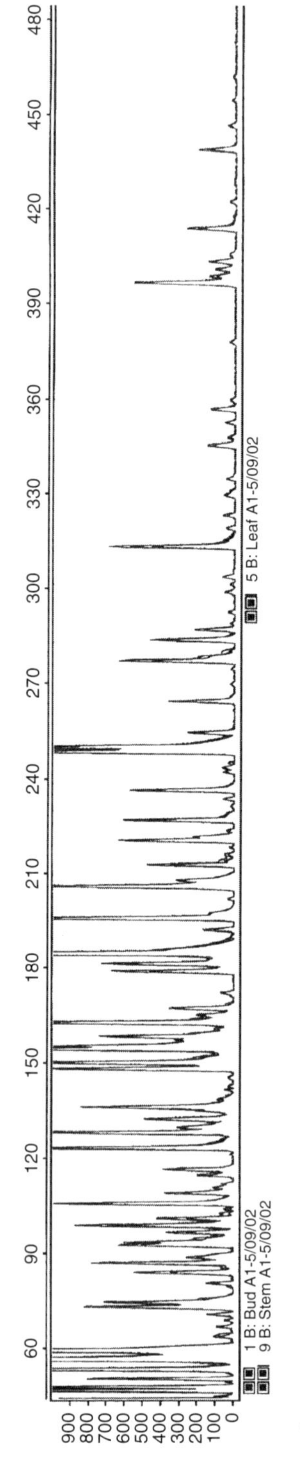

Figure 12.1 B) DNA extracted from different tissue sources of the same plant sample (stem, bud, and leaf) yield comparable amplified fragment length polymorphism (AFLP) profiles. X-axis, size in nucleotides; y-axis, relative fluorescence units (RFU).

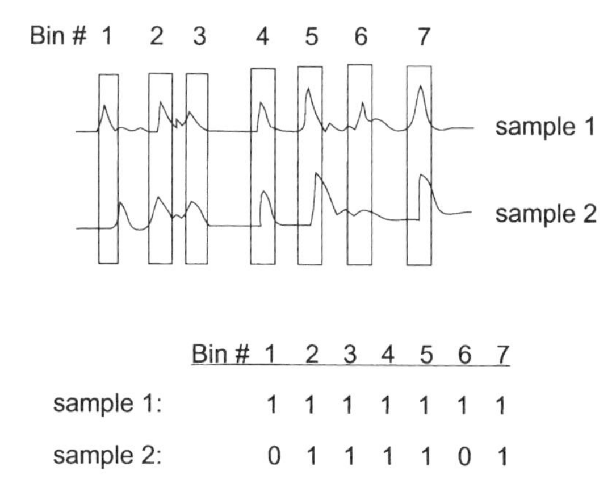

Figure 12.2 An illustration of how Genotyper software can be useful for defining categories for scoring peaks found to be variable between different samples. If a peak is present, it is given a value of 1; if absent, a value of 0. In this manner, the variable peaks within an AFLP profile can be converted to a data format that can be used to create a database.

12.3.3 Unrelated Marijuana Samples Have Different AFLP Profiles

Nineteen different marijuana seizure samples obtained from adjudicated cases from the Vermont Crime Laboratory were processed to obtain AFLP profiles and the results compared. The Vermont marijuana samples were typical of street seizure material that has been observed in Connecticut and other New England states. These samples consisted of mixtures of leaves, stems, seeds, and flower buds. Individual fragments were selected for processing to avoid the possibility of typing plants from two different sources. The extracted DNA from these seizure samples was partially degraded but met the criteria for AFLP processing listed in the materials and methods. Variable categories have been selected for scoring the AFLP profiles generated by PCR primer set A1 to binary code (see Figure 12.2). When comparing the AFLP profiles, presence (scored as 1) or absence (scored as 0) of a peak is easily determined by visual examination of the Genotyper results. A total of 100 variable AFLP markers are scored for the two PCR primer sets. Three positive control peaks are present in the profiles generated by each primer set and are useful for assessing the data on a comparable relative fluorescence (RFU) scale for comparing two or more profiles.

12.4 Discussion

Our data show that AFLP analysis is a useful method for establishing genetic profiles from seized marijuana samples. These profiles can be utilized to provide genetic evidence that clonally propagated marijuana samples are being shared between two locations or have a common origin. This information can provide an investigative lead for law enforcement when tracking common sources of marijuana and for linking source, distributor, and user. AFLP analysis performed on floral bud, stem, and fresh and dried leaf tissue was successful in generating interpretable profiles. Interestingly, all of the Vermont seizure samples had different AFLP profiles, suggesting they may be genetically unrelated or distantly related. Although it is thought that most marijuana cultivars are highly inbred and derived from a few superior genetic lines, AFLP markers are able to distinguish sufficient genetic diversity within the marijuana samples we have surveyed to date. Based on our current data, two AFLP primer sets (and frequently a single primer set) are sufficient to determine that two marijuana samples are not clonal in nature. AFLP analysis is a promising new method for forensic analyses of single-source plant material and has the potential to be applied to other forms of nonhuman biological evidence and additional types of casework.

Acknowledgments

The authors would like to thank Dr. Eric Buel, the Vermont Crime Laboratory, Captain Peter Warren, and members of the Connecticut Statewide Narcotics Task Force for their assistance in providing either samples or helpful information. A large number of students worked on the AFLP research project for marijuana and their assistance was greatly appreciated: Eric Carita, Nicholas CS Yang, Elizabeth McClure Baker, Joselle Germano, Edward Jachimowicz III, Shane Lumpkins, Khartika Divarkaran, and Michelle Irvin.

References

1. Pollan, M., *The Botany of Desire: A Plant's-Eye View of the World*, New York: Random House, 2001.
2. Clarke, R.C., *Marijuana Botany — An Advanced Study: The Propagation and Breeding of Distinctive Cannabis*, Berkeley, CA: RONIN Publishing, 1981.
3. Frank, M. and Rosenthal, E., *Marijuana Grower's Guide*, Los Angeles, CA: Red Eye Press, 1990.
4. Vos, P. et al., AFLP: a new technique for DNA fingerprinting, *Nucleic Acids Research*, 23, 4407, 1995.
5. Becker, J. et al., Combined mapping of AFLP and RFLP markers in barley, *Mol. Gen. Genet.*, 249, 65, 1995.
6. Van Eck, H.J. et al., The inheritance and chromosomal location of AFLP markers in a non-inbred potato offspring, *Molecular Breeding*, 1, 397, 1995.

7. Thomas, C.M. et al., Identification of amplified restriction fragment polymorphism (AFLP) markers tightly linked to the tomato cf.-9 gene for resistance to *Cladosporium fulvum*, *Plant J.*, 8, 785, 1995.
8. Lin, J-J et al., Identification of molecular markers in soybean—comparing RFLP, RAPD, and AFLP DNA mapping techniques, *Plant Mol. Biol. Rep.*, 14, 156, 1996.
9. Menz, M. et al., A high-density genetic map of *Sorghum bicolor* (L.) Moench based on 2926 AFLP, RFLP and SSR markers, *Plant Mol. Biol.*, 48, 483, 2002.
10. Peters, J. et al., A physical amplified fragment-length polymorphism map of *Arabidopsis*, *Plant Physiol.*, 127, 1579, 2001.
11. Jakse, J., Kindlhofer, K., and Javornik, B., Assessment of genetic variation and differentiation of hop genotypes by microsatellite and AFLP markers, *Genome*, 44, 883, 2001.
12. Ren, N. and Timko, M., AFLP analysis of genetic polymorphism and evolutionary relationships among cultivated and wild *Nicotiana* species, *Genome*, 44, 559, 2001.
13. Loh, J. et al., Amplified fragment length polymorphism fingerprinting of 16 banana cultivars (*Musa* cvs.), *Mol. Phylogenet. Evol.*, 17, 360, 2000.
14. Suyama, Y., Obayashi, K., and Hayashi, I., Clonal structure in a dwarf bamboo (*Sasa senanensis*) population inferred from amplified fragment length polymorphism (AFLP) fingerprints, *Mol. Ecol.*, 9, 901, 2000.
15. Bensch, S. et al., Amplified fragment length polymorphism analysis identifies hybrids between two subspecies of warblers, *Mol. Ecol.*, 11, 473, 2002.
16. Liu, Z. et al., Inheritance and usefulness of AFLP markers in channel catfish (*Ictalurus punctatus*), blue catfish (*I. Furcatus*) and their F1, F2 and backcross hybrids, *Mol. Gen. Genet.*, 258, 260, 1998.
17. Parsons, Y. and Shaw, K., Species boundaries and genetic diversity among Hawaiian crickets of the genus *Laupala* identified using amplified fragment length polymorphism, *Mol. Ecol.*, 10, 1765, 2001.
18. Weeks, A., van Opijnen, T., and Breeuwer, J., AFLP fingerprinting for assessing intraspecific variation and genome mapping in mites, *Exp. Appl. Acarol.*, 24, 775, 2000.
19. Zhong, S. et al., A molecular genetic map and electrophoretic karyotype of the plant pathogenic fungus *Cochliobolus sativus*, *Mol. Plant Microbe Interact.*, 15, 481, 2002.
20. Janssen, P. et al., Evaluation of the DNA fingerprinting method AFLP as a new tool in bacterial taxonomy, *Microbiology*, 142, 1881, 1996.
21. Dijkshoorn, L. et al., Comparison of outbreak and nonoutbreak *Acinetobacter baumannii* strains by genotypic and phenotypic methods, *J. Clin. Microbiol.*, 34, 1519, 1996.
22. Avrova, A. et al., Application of amplified fragment length polymorphism fingerprinting for taxonomy and identification of the soft rot bacteria *Erwinia carotovora* and *Erwinia chrysanthemi*, *Appl. Environ. Microbiol.*, 68, 1499, 2002.

23. Geornaras, I. et al., Amplified fragment length polymorphism fingerprinting of *Pseudomonas* strains from a poultry processing plant, *Applied Environ. Microbiol.*, 65, 3828, 1999.
24. Miller Coyle, H. et al., The green revolution: botanical contributions to forensics and drug enforcement, *Croatian Medical Journal*, 42, 340, 2001.
25. Miller Coyle, H. et al., A simple DNA extraction method for *Cannabis sativa* samples useful for AFLP analysis, *J. Forensic Sci.*, 2003 (in press).
26. Lee, H.C. et al., DNA typing in forensic science: I. Theory and background, *American Journal of Forensic Medicine and Pathology*, 15, 269, 1995.
27. Bever, R. and Cimino, M., Forensic molecular botany: identification of plants from trace evidence, Twelfth International Symposium on Human Identification, Biloxi, MS; 2001.
28. Nakamura, G.R., Forensic aspects of cystolith hairs of *Cannabis* and other plants, *J. Assoc. Off. Anal. Chem.*, 52, 5, 1969.
29. Butler, W., Duquenois-Levine test for marijuana, *J. Assoc. Off. Anal. Chem.*, 45, 597, 1962.
30. Pitt, C.G., Hendron, R.W., and Hsia, R.S., The specificity of the Duquenois color test for marijuana and hashish, *J. Forensic Sci.*, 17, 693, 1972.
31. Barni Comparini, I. and Centini, F., Packed column chromatography, high resolution gas chromatography and high pressure liquid chromatography in comparison for the analysis of *Cannabis* constituents, *Forensic Sci. Int.*, 21, 129, 1983.
32. Siniscalco Gigliano, G., Preliminary data on the usefulness of internal transcribed spacer I (ITS1) sequence in *Cannabis sativa* L. identification, *J. Forensic Sci.*, 44, 475, 1999.
33. Siniscalco Gigliano, G., Caputo, P., and Cozzolino, S., Ribosomal DNA analysis as a tool for the identification of *Cannabis sativa* L. specimens of forensic interest, *Science & Justice*, 37, 171, 1997.

Legal Considerations for Acceptance of New Forensic Methods in Court

13

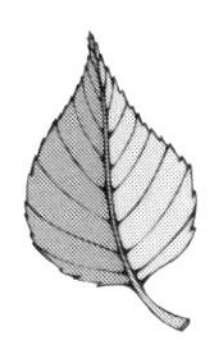

TIMOTHY PALMBACH[1] AND GARY SHUTLER[2]

Contents

[1] Sections 13.1–13.5, 13.7, 13.8.
[2] Section 13.6.

0-8493-1529-8/05/$0.00+$1.50

13.1 Introduction

DNA profiles obtained from a forensic case sample can exclude an individual as a possible source of the sample or include this individual as a potential donor, often with a high degree of discrimination. Initially, courts responded cautiously to the powerful individualizing abilities of DNA, recognizing the complexity of DNA typing procedures and the difficulty of presenting such results to a jury. However, a plethora of research, evaluation, and technical advancement as well as the education of the court have transformed the attitudes of judge and jury into a level of general understanding and acceptance for human DNA typing methods.

13.2 Early Issues

Some of the early apprehensions and roadblocks to the admissibility of DNA evidence involved general concerns with DNA typing procedures and methodologies. These types of contentions developed into inquiries relating to the proper application of accepted or standard methodologies to the specific case in question. In addition, almost from the onset and, still today, many of the challenges came in reference to interpretation of DNA typing results. Many of these arguments were formulated around statistical applications presented to render some "meaning" to the derived profiles. Originally, there was some debate over whether the presentation of a DNA match by itself (i.e., sans statistical random match frequencies) was overly prejudicial. Currently, most of these matters have been resoundly addressed by courts throughout the world, and it is no longer common for a legal challenge to succeed on the basic issues, procedures, or methodologies. However, as research and development continues to expand our understanding and technical capabilities in the forensic DNA community, admissibility issues will persist for new technologies and their novel applications.

Prior to 1986 and the introduction of "DNA fingerprinting," traditional serological methods were being used in forensic casework to analyze biological samples. These methods analyzed the Human Leucocyte Antigen (HLA) system, components of blood involving cellular antigens, cellular and serum proteins and isoenzymes, as well as various components of other body fluids or tissues. In 1986, Dr. Alec Jeffreys at the U.K.'s Leicester University applied his novel DNA typing methods to resolve evidence in a criminal investigation. This case involved two separate rape–murder investigations. Lynda Mann, one of the victims, was murdered on November 22, 1983. The second victim, Dawn Ashworth, was found on August 22, 1986. Following Dr. Jeffreys' proposal, the police requested that all men living in the areas of these murders submit a blood sample for DNA testing. These samples would be compared to DNA profiles obtained from the analysis of semen samples recovered from the two victims. During this process the ultimate defendant, Colin Pitchfork, paid someone to impersonate him and provide his blood sample in place of Pitchfork's sample. On learning of this switch, the police located Pitchfork and took a sample of his blood. Use of Jeffreys' novel DNA fingerprinting method determined a match between the semen sample and Pitchfork's known blood sample.[1]

DNA typing was quickly integrated in the U.S. and began providing valuable information. By 1990, more than two thousand courts in the U.S., in 49 states and the District of Columbia, had successfully used DNA evidence in criminal matters.[2] Despite its rapid integration and widespread use, many courts in the late 1980s and early '90s were conducting full-scale admissibility hearings on DNA evidence. During this time period, the predominant threshold admissibility standard was that established in Frye.[3] An application of a Frye standard typically involved an inquiry to determine if there was a general acceptance of the theory underlying DNA typing, and that the specific techniques used to derive the results were generally accepted in the relevant scientific communities. During these judicial inquiries there were common threads of concern. Courts wanted to ensure quality assurance of testing procedures, including validation of those procedures, and of the individual laboratories and examiners conducting the testing. Also, there was general concern over methods used to calculate frequencies of occurrence with inclusion cases. Moreover, questions and concerns arose about the application of population genetics and assumptions associated with those interpretations.

One of the early comprehensive hearings on DNA evidence was held in New York State, *People v. Joseph Castro.*[4] In *Castro*, prosecutors sought to show that bloodstains on the defendant's wristwatch were from the victim, and not from the defendant, despite his claims to the contrary. After a lengthy admissibility hearing, the Castro court held the following:

- There is a general scientific acceptance of the theory underlying DNA identification.
- DNA forensic identification techniques and experiments are generally accepted in the scientific community and can produce reliable results; hence, the Frye standard is satisfied.
- A pretrial hearing should be conducted to determine if the testing laboratory substantially performed the scientifically accepted tests and techniques, yielding sufficiently reliable results to be admissible as a question of fact for the jury.

However, the Castro court ultimately ruled the DNA evidence as inadmissible because of the unreliability of some of the applied statistical calculations, and the utilized size or ratio of the population frequency. Reacting to concerns raised in Castro, as well as many other adjudicated cases involving DNA evidence, several research and advisory groups were commissioned. In 1988, the Office of Technology Assessment, Congressional Committee, published a preliminary report. This report was published in a book titled *Genetic Witness — Forensic Uses of DNA Tests.* In 1992 and in subsequent years, the Committee on DNA Technology in Forensic Science, National Research Council, published its commonly cited findings. These reports served as a resource for courts to rely on as they continued to grapple with the complex issues associated with DNA evidence.

The committee's report was published in 1992 by National Academy Press. In response to apparent misrepresentations of the report in news articles, the committee issued the following brief overview statement: "We recommend that the use of DNA evidence analysis for forensic purposes, including the resolution of both criminal and civil cases, be continued while improvements and changes suggested in this report are being made. There is no need for a general moratorium on the use of DNA typing results either in investigation or in the courts. We regard the accreditation and proficiency testing of DNA typing laboratories as essential to the scientific accuracy, reliability, and acceptability of DNA typing evidence in the future. Laboratories involved in forensic DNA typing should move quickly to establish quality assurance programs. After a sufficient time for implementation of quality assurance programs has passed, courts should view quality control as necessary for general acceptance."

13.3 General Admissibility

Overall, the use and general acceptance of DNA evidence in court has become the norm, with only an occasional case-specific issue. Precedence and a history of extensive judicial scrutiny, coupled with confidence won by the relevant scientific communities responding to early criticisms, account for the relative lack of objections relating to the validity of the basic principles and methodology associated with DNA typing. However, this status came only after years of scrutiny and refinement. Further, as research and development provide us with new methodologies or applications, foundational work will once again be required before the offered DNA results will be admitted into evidence. Testimony derived from scientific or technical testing must meet a threshold admissibility standard, and any form of DNA testing is subject to this form of judicial review.

13.3.1 Frye

Historically, and even in some jurisdictions today, the base standard was established in the 1923 case of *Frye v. United States.*[5] In Frye, the Court of Appeals for the District of Columbia rejected the defendant's motion to introduce polygraph evidence, holding, "Just when a scientific principle or discovery crosses the line between the experimental and demonstrable stages is difficult to define. Somewhere in this twilight zone, the evidential force of the principle must be recognized, and while courts will go a long way in admitting expert testimony deduced from a well-recognized scientific principle or discovery, the thing from which the deduction is made must be sufficiently established to have gained general acceptance in the particular field it belongs." This "general acceptance" model was almost universally adopted as the sole standard for admissibility of scientific evidence for approximately 70 years.

As courts grappled with issues associated with the Frye standard, such as what aspect of the scientific principle or testing must be shown to be generally accepted, the Frye standard was modified or clarified. A multiprong test was developed. Commonly, there were three prongs articulated in this refined model. Frye-1 evaluates whether the test was based on a fundamental scientific principle or discovery. Frye-2 determines if the specific technique used in the analysis was also based on fundamental scientific principles or discoveries. Finally, Frye-3 examines a technique-specific application, in the specific case before the court, on which the expert testimony is to be based.[6]

A majority of states adopted the first two prongs; however, only a handful adopted Frye-3.

Applying the Frye standard of admissibility to proffered DNA testimony raises several issues. First is the strict application of Frye, which tends to address only the general acceptance of the underlying scientific principles. While trial judges undoubtedly consider methodology and the application of the principle in the case at hand, Frye does not require that type of inquiry. A second issue involves the difficulty in determining if there are separate distinct components of a scientific analysis, which each require a separate Frye analysis. With DNA testimony, several distinct fields encompass the overall procedures and subsequent testimony. One portion involves the application of statistics to the data derived from DNA typing; others involve human genetics, microbiology, or other related scientific disciplines. This is a relevant matter in that differences exist among different scientific and mathematic communities. For example, genetics and statistics experts have disagreed as to the proper statistical interpretation. Third, a Frye inquiry may be initiated to focus on the application of a generally accepted technique in that specific case. This is often viewed as a Frye-3 analysis. Many courts accepting a need to conduct this type of inquiry cited it as a reason why Frye is not the appropriate standard.

13.3.2 Daubert

As the reliance on scientific evidence increased, courts began to sway from the restrictive general acceptance protocol of Frye. Legislatures and courts sought a solution to growing feelings of disdain with Frye. Then in 1975, the Federal Rules of Evidence (FRE) were adopted with a new admissibility standard. The newly adopted Federal Rules addressing opinions and expert testimony are found within Article VII — Opinions and Expert Testimony. Rules 702 and 703 directly address issues of admissibility for scientific evidence.

13.3.3 FRE 702 and FRE 703

FRE 702 states, "If scientific, technical, or other specialized knowledge will assist the trier of fact to understand the evidence or to determine a fact in issue, a witness qualified as an expert by knowledge, skill, experience, training, or education, may testify thereto in the form of an opinion or otherwise." According to FRE 703, "The facts or data in a particular case upon which an expert bases an opinion or inferences may be those perceived by or made known to the expert at or before the hearing, if of a type reasonably relied upon by experts in the particular field in forming opinions or inferences upon the subject, the facts or data need not be admissible in evidence."

Comparing FRE to Frye, FRE is generally less restrictive. Under FRE, judges are allowed to consider a variety of factors, including but not limited to general acceptance. These new rules were intended, in part, to remedy what was seen as a tendency of common law evidence rules, which elevate form over substance, in determining the admissibility of evidence.[7] These new rules allowed a judge to render her decision on a case-by-case subjective analysis. Thus, admissibility was not viewed with a black-and-white procedural perspective. In contrast, Frye was viewed by many as an overly conservative objective standard in that the entire issue of admissibility was based on nothing more than the general acceptance of the underlying scientific principle.

Following the adoption of FRE, courts acknowledged a conflict between the Federal Rules of Evidence and the Frye standard.[8] Prior to a major shift in the 1993 Daubert ruling, 20 states had rejected Frye and adopted a "helpfulness" or "relevance" test for the admissibility of scientific evidence.[9] In 1993, the U.S. Supreme Court addressed this apparent conflict in the case of *Daubert v. Merrell Pharmaceuticals.*[10] The Court, in Daubert, explicitly held that Federal Rules of Evidence 702 supersedes the Frye "general acceptance" standard.

Daubert articulated several factors to be considered when determining the admissibility of a new method:

- Whether a theory or technique in question can be and has been tested
- Whether it has been subjected to peer review and publication

- Its known error rate
- The existence and maintenance of standards controlling its operation
- Whether it has attracted widespread acceptance within a relevant scientific community

The inquiry is a flexible one, and its focus must be solely on principles and methodology, not on the conclusions they generate. It is flexible in that a judge has the discretion of apportioning the weight to be given to any of the listed factors. An outcome of Daubert is that the presiding judge is the "gatekeeper." Proponents of Daubert view this shift as a proper means of reallocating authority away from scientists back to the trial judge. No longer is a judge bound by an expert witness' testimony establishing the procedures and principles relied on as being generally accepted in the relevant scientific community. Rather, a judge assumes the responsibility of evaluating scientific methods herself, and determining if those methods are relevant and reliable. However, as some critics have declared, this judicial gatekeeping function can, in some instances, be a problem associated with Daubert as a scientifically sound procedure may be incorrectly excluded due to the judge's lack of understanding of the scientific principles. Moreover, taken to the opposite extreme, a judge may inadvertently admit evidence based on "junk science."

Since Daubert, numerous state courts have enacted their own admissibility standards consistent with Daubert. For example, in Connecticut, the Court may consider several criteria when deciding whether the underlying scientific theories or techniques are scientifically reliable. These criteria include

- General acceptance
- Whether the method has been tested and subject to peer review
- Known or potential error rates
- Prestige and background of the expert
- Subjective interpretation versus objectively verifiable criteria
- Ability of the expert to explain the data in a manner so that the fact finder can draw its own conclusions
- Whether the technique was developed solely for the case at hand[11]

As with any new provision there was testing and refinement in years following the enactment of Daubert standards. The U.S. Supreme Court made modifications primarily in two cases that followed Daubert. In the *Joiner* case, the court held that the reviewing court must apply an abuse-of-discretion standard when it reviews the trial court's decision to admit or exclude expert testimony. That standard applies as much to the trial court's decisions about how to determine reliability as to its ultimate conclusion. Thus, whether Daubert's specific factors are, or are not, reasonable measures of reliability in a particular case are a matter that the law grants the trial judge broad latitude to determine.[12] In the second modifying case, *Kumho Tire*, the Supreme Court upheld a trial judge's decision to exclude technical testimony. Specifically, the technical testimony was from an engineer who testified about tire failures. This ruling clarified that the Daubert standard applied to all technical and specialized knowledge, not merely scientific testimony. This decision rejected any presumption that technical knowledge is reliable by its nature or because it is based on proven scientific principles.[13]

Responding to the judicial modifications, the Congress approved proposed changes to Federal Rules of Evidence for Rule 701 — Opinion Testimony by Lay Witnesses, Rule 702 — Testimony by Experts, and Rule 703 — Bases of Opinion Testimony by Experts. In the following sections, we discuss the additions made to each of these rules (the underlined text indicates the 2000 modification).

13.3.4 FRE 701

FRE 701 specifies: "if the witness is not testifying as an expert, the witness' testimony in the form of opinions or inferences is limited to those opinions or inferences which are (a) rationally based on the perception of the witness, and (b) helpful to a clear understanding of the witness's testimony or the determination of fact in issue, and (c) not based on scientific, technical, or other specialized knowledge within the scope of Rule 702."

13.3.5 FRE 702

If scientific, technical, or other specialized knowledge will assist the trier of fact to understand the evidence or to determine a fact in issue, a witness qualified as an expert by knowledge, skill, experience, training, or education, may testify thereto in the form of an opinion or otherwise, if (1) the testimony is based upon sufficient facts or data, (2) the testimony is the product of reliable principles and methods, and (3) the witness has applied the principles and methods reliably to the facts of the case.

13.3.6 FRE 703

The facts or data in a particular case upon which an expert bases an opinion or inferences may be those perceived by or made known to the expert at or before the hearing, if of a type reasonably relied upon by experts in the particular field in forming opinions or inferences upon the subject, the facts or data need not be admissible in evidence in order for the opinion or inferences to be admitted. Facts or data that are otherwise inadmissible shall not be disclosed to the jury by the proponent of the opinion or inferences unless the court determines that their probative value in assisting the jury to evaluate the expert's opinion substantially outweighs their prejudicial effect.

Yet despite changes that appear consistent with the holdings in *Daubert*, *Joiner*, and *Kumho* there will still be occasional challenges and seemingly inconsistent rulings in a specific case. Some legal scholars believe that these changes will leave much room for judicial creativity in implementing them.[14]

13.4 Applying These Standards to Nonhuman DNA Analysis

Nonhuman DNA testing has already been successfully incorporated into forensic case examinations and in several key cases been admitted into evidence. These nonhuman DNA applications involved the analysis of animal hair, botanical specimens, and bacterial or microbiological samples. Laboratories that are conducting nonhuman DNA typing have modified established, reliable, and generally accepted protocols to conduct their analyses. Courts responding to these examinations relied on testimony establishing that these new methods were reliable and met those same admissibility standards that human DNA typing had been subjected to in their particular jurisdiction.

Sensabaugh and Kaye examine the important issues that the courts should use in evaluating the admissibility of novel DNA evidence. The factors they propose are based on precedents already established for human DNA identification. These factors include

- The novelty of the application
- The validity of the underlying scientific theory
- The validity of any statistical interpretations
- The availability of a relevant scientific community to consult in assessing the application

As with any of the new advancements in DNA typing technologies, standards and admissibility criteria will need to be met through intensive research and validation studies prior to everyday implementation for forensic casework. Since the first use of DNA typing technology in the U.K., the concept of identification by DNA testing has become common knowledge. Although some samples remain difficult to type, there is no longer a novelty associated with using DNA analysis to identify humans. After decades of legal challenges, the scientific validity of human DNA testing has been well established. The DNA typing techniques used for human studies applies as well to the study of nonhuman DNA. Yet, nonhuman samples have not been viewed with the same level of acceptance as with results generated from testing of human DNA. An examination of court cases involving the use of nonhuman DNA reveals several consistent key judicial concerns. These concerns are often not with the validity of the underlying methodology or the application of those techniques to a particular nonhuman DNA sample; rather they involve the interpretation or statistical analysis applied to the results. When presenting nonhuman evidence in court, several possible conclusions regarding testing can be reached. The known reference sample may be excluded as the donor of the evidentiary sample. The results may be indeterminate due to a lack of a result or inconclusive due to a weak result (partial DNA profile). Alternatively, the known reference sample may be consistent with being the donor of the sample, and if included, the match may be stated as such or a statistical value may be assigned to it. One of the most controversial aspects of DNA is the assignment of a statistical value and the population database utilized to reach an appropriate estimate for a random match probability.

Most often a major issue is that specific loci or alleles studied have not been adequately researched and validated to ensure that that they meet all of the necessary forensic criteria, such as studies demonstrating that the derived databases are in Hardy–Weinberg and linkage equilibrium. Issues or concerns involving contamination, matching criteria, or heteroplasmy (in the case of mitochondrial DNA testing) tend to go to the weight attributed to the nonhuman DNA evidence rather than to a question of its admissibility at trial. As a result, courts either disallow the admission of nonhuman DNA evidence or admit the results into evidence but instruct the jury to give it appropriate limited weight.

Those instances where the proffered DNA evidence was not admitted generally involve the court's findings that studies available on nonhuman DNA polymorphic loci, or their relative frequency of occurrence, are insufficient for specific markers to be accurate forensic indicators. Then, the court views the limited probative value of those nonhuman DNA results as not outweighing the undue prejudice.[15] Cases in which nonhuman DNA testing results were offered, but with a valid and reasonable match statement, tend to be more readily admitted into evidence.

13.5 Lineage-Based Testing

There are significant differences in the manner in which organisms inherit different forms of DNA. Nuclear DNA is inherited after genetic recombination in a child as 50% from each parent, maternal and paternal. However, mitochondrial DNA (maternal lineage) and the male (Y) sex chromosome (paternal lineage) are inherited from a single parent and, with the exception of mutation, maintained in the same state from generation to generation. These genetically defined differences in inheritance patterns result in related individuals having the same form of mitochondrial or Y-chromosome DNA, and this affects the statistical assessment for random match probability calculations. In addition, these forms of DNA testing (whether they are for human or nonhuman DNA samples) do not have the same discrimination ability of nuclear DNA testing because a certain percentage of unrelated individuals in the population are expected to share the same form of DNA (e.g., haplotype sharing). The populations that are constructed to provide estimates for the percentage of haplotype sharing for these forms of testing will need to be carefully considered to ascertain whether they are representative of larger populations. However, these accepted forms of human lineage-based DNA tests will set a standard for DNA testing of populations of other organisms that have combined forms of genetic inheritance (i.e., random mating and asexual reproduction/cloning).

A case in point involved a robbery investigation where police recovered two head hairs from a sweatshirt located alongside a road leading away from the robbery scene. The FBI laboratory performed human mitochondrial DNA analysis and compared profiles generated from these two hairs to the defendant's head hairs. As a result of this analysis, it was reported that the defendant could not be excluded as the source of questioned samples.[16] During the trial, a mitochondrial DNA expert provided testimony to properly define and limit the interpretation of the results. This expert explained that because mitochondrial DNA has only one marker, the probability of a random match is much higher between mitochondrial DNA samples than between nuclear DNA samples. Moreover, this witness stated that the examiner could not positively establish identity on the basis of mitochondrial DNA because all those having a common maternal lineage, absent mutation, share the same mitochondrial DNA. Finally, this expert witness informed the court that he had derived his statistical numbers relevant to the match based on the conservative counting method. With this method, he estimated the rarity or prevalence of a given mitochondrial DNA sequence based on whether the sequence had been observed in the database and, if so, how often it had been observed. Based on these limitations and clarifications provided to the trial court, the court ruled that the testimony was statistically sound and that it was likely to be helpful to the jury in assessing the probative value of the mitochondrial DNA evidence. However, during jury instructions the judge gave the following limiting instruction: "You are to consider the mitochondrial DNA evidence in this case giving it the weight that you deem appropriate. However, you must remember that nuclear DNA evidence is entirely different and nuclear DNA evidence is not involved in this case."

Other jurisdictions, including some of those still relying on a Frye-based standard rather than a Daubert admissibility standard, also allowed for acceptance of newer DNA technologies after the trial court determined that the offered evidence was of the type generally accepted in the scientific community.

In 2003, a Michigan court allowed for evidence derived from mitochondrial DNA testing.[17] At issue in that case was the mitochondrial DNA testing conducted on three hairs. Two hairs subsequently determined to belong to the defendant were found at the crime scene. The third hair, identified as belonging to the victim, was found in the defendant's bedroom. Specifically the court dealt with three DNA-related issues. First, as the basic matter, a determination of whether mitochondrial DNA testing has been generally accepted in the scientific community was made. Second, a claim that generally accepted laboratory procedures were not followed in this case was addressed. Finally, it was considered if the court erred in allowing the use of the "counting method" in describing the mitochondrial DNA results. Relative to the statistical issue, the court relied on expert testimony that the counting method, which merely reports how many times a particular sequence has been seen before in the FBI database, is "excessively conservative" and favors the defendant. Ultimately, this court held that the statistical analysis of DNA testing is relevant to the weight of the evidence, not its admissibility.

Applying the Davis–Frye test, the court held that such evidence was generally accepted in the scientific community. Their holding was based on testimony from various scientists and academicians who had published heavily in the field and the long-standing use of such evidence in criminal cases in Europe. Consistent with a vast majority of other courts, this court declared that problems in the procedures followed by the two laboratories that performed DNA testing were very unlikely to have affected the test results and, therefore, went to the weight of the evidence, not its admissibility.

13.6 Landmark Cases for Nonhuman DNA Typing

13.6.1 Plant Evidence

The first murder case involving an application of plant DNA typing occurred for a 1992 case in Arizona.[18] A young woman's body was found strangled in the desert near several paloverde trees. A pager near the body belonging to the owner of a white pickup truck seen in the area at about the time the murder would have occurred led to the identification of a suspect. Two paloverde seedpods were found in a search of the vehicle. Recent damage on one of the paloverde trees at the crime scene led detectives to speculate it was caused by the suspect's truck. A randomly amplified polymorphic DNA (RAPD) method was used by Dr. Timothy Helentjaris from the University of Arizona to compare the DNA profile from the seedpods to the suspect tree. In addition, he sampled other paloverde trees collected from around Maricopa County (9 trees at the scene and 19 from other locations). He concluded in a blind study that the seedpods came from the damaged tree at the crime scene. This evidence tied the truck to the scene and helped to resolve the case in court. The suspect was found to be guilty at the trial.

On appeal of the paloverde case, the trial court's finding that RAPD technology is generally accepted in the scientific community (the Frye Standard) and the subsequent match testimony offered by the DNA analyst was affirmed. At trial, Dr. Helentjaris testified that the odds of a random match in this case were one in a million. An expert in opposition stated that the chance of a random match was one in 136,000. Regarding the statistical issue, the reviewing court concluded that the soundest approach was to permit expert testimony on the ultimate scientific issue at hand: whether one sample is the genetic match of another. This court felt that opponents of DNA match testimony are free to challenge its foundation and introduce controverting evidence.

Although plant DNA evidence (rather than morphological comparisons) has been rarely utilized to date in criminal cases, it has been used fairly extensively in civil cases. These cases typically center on patent infringement for crop cultivation. Plant patents are designed to provide incentive to seed companies and horticulturalists to develop improved crop and ornamental varieties of plants. The patent of a new variety then prohibits the use or sale of that variety, thereby eliminating competition and allowing the plant breeder to recoup some economic proceeds for his efforts. Typically, plant patents extend for a 20-year period. There are three distinct but complementary forms of intellectual property protection for agricultural innovations:

- A utility patent for technical invention
- A plant patent for vegetatively propagated plants (plants grown from cuttings rather than seed)
- The Plant Variety Protection Certificate (for plants grown from seed)

In a landmark case, in October 2001, the Supreme Court held a hearing for the case of *J.E.M. Agricultural Supply v. Pioneer Hi-Bred.* Pioneer Hi-Bred filed the original suit for J.E.M. Supply selling hybrid corn seeds patented by Pioneer Hi-Bred. J.E.M. Agricultural Supply countersued and the case went forward to the Supreme Court. At issue was the inherent right for an individual or corporation to patent living material. Through bioengineering or traditional breeding, a living plant may be significantly altered or improved and the new trait is then considered patentable. Ultimately, the Supreme Court held that utility patents could be issued for plants even though distinct protections were also available under the Plant Variety Protection Act and the Plant Patent Act. With appropriate research and validation, it is anticipated that plant DNA typing techniques used in the civil courts may be transferable to botanical evidence associated with criminal cases.

Domesticated animal hairs have been a common source of nonhuman DNA samples in which DNA analysis is sought. Most often these samples are in the form of animal hairs located at a crime scene, or as trace evidence on a victim, suspect, or object associated with the crime. Generally, there is a request to link the animal hairs seized during the investigation with known hairs from either the victim's or suspect's pet, or from a source other than a known pet that has potential relevance in that it can establish associations. The two major necessary initiatives to provide these services were to develop and validate a procedure that would generate a DNA profile from the animal hair, and then to create a relevant database to compare with the derived profile and develop statistical matches. Over the years specific methodologies and databases were developed and refined. In the 1990s, a domestic dog database was developed using sequence analysis of mitochondrial DNA.[19] During the same time period a domestic feline database was generated using STR loci.[20]

13.6.2 Cat Hair Evidence

In a 1994 landmark case on Prince Edward Island, animal DNA played an important role in a murder case. A woman had disappeared and her body was located in a shallow grave in a heavily wooded area. Her ex-husband was the prime suspect. Investigators found a brown leather jacket near the recovered body. Bloodstains on the jacket had a DNA profile that matched the victim. In addition to witness statements, two white cat hairs were collected from the garment. The suspect lived with his parents and they owned a white cat called Snowball. The hair was sent to the laboratory involved with the National Cancer Institute's Cat Genome Project in Maryland. Dr. Marilyn Menotti-Raymond's laboratory used ten STR loci for typing and found a DNA profile match to the suspect family's cat. This novel nonhuman evidence was used at trial and the defendant was found guilty of murder.

13.6.3 Dog Evidence

Two West Coast cases of murder (1991 and 1996, respectively) and one animal attack in Oklahoma (1999) were resolved with dog DNA typing from Dr. Joy Halverson's California laboratory. The first involved a man and his dog, both found dead with head wounds in Vernon, British Columbia. Clothing recovered from a trash bin had human bloodstains of mixed origin with the major component of the STR DNA profile matching the murdered man and a minor partial DNA profile consistent with coming from the suspect. Other blood from the clothing was shown to be from the dog and matched at 11 STR loci to the victim's terrier. This evidence aided in resolving the case into a guilty plea. The second case was a homicide and a charge of animal cruelty. Canine bloodstains from the accused's clothing yielded an STR profile that matched the victim's deceased dog, Chief. This evidence was accepted in a Seattle, WA, court for a guilty verdict. In Tulsa, OK, an animal attack involved a 74-year-old woman mauled by a bit bull terrier. The DNA profile generated from dog saliva on the victim's clothing matched the suspect dog. The impounded dog died of a tumor prior to being destroyed per judge's order.

13.7 Further Court Cases Addressing Concerns with Nonhuman DNA

In 2001, the U.S. Court of Appeals for the Eighth Circuit heard a case involving a claim by the defendant, Dr. Boswell, that the district court has abused its discretion by admitting into evidence DNA test results taken from swine serum samples submitted for testing.[21] This testing was done in an attempt to confirm Dr. Boswell's claim that certain samples came from specific swine. The swine DNA was tested using a polymerase chain reaction (PCR) method.

The appeals court had to determine whether or not the district court made a clear error of judgment in weighing the facts on the basis of the record before it. Citing Daubert, the court stated that the admission of scientific expert testimony is dependent on the court's determination that the proposed testimony constitutes scientific knowledge that will assist the trier of fact to understand or determine a fact in issue. After reviewing the trial record, this court determined that there was sufficient evidence that the district court was justified in permitting the admission of PCR test results of the swine DNA.

The basis for the court's holding was the following articulated reasons. First, the PCR process is approximately ten years old and has undergone extensive testing. Numerous courts have recognized and endorsed the use of PCR analysis for forensic purposes. PCR analysis is an accurate method of analysis when a protocol that conforms to guidelines accepted by members of the forensic community is followed and controls are employed. Finally, citing one of their previous decisions allowing for the admissibility of PCR-based DNA analysis, the court held that any alleged deficiencies must so alter the PCR methodology as to make the test results inadmissible.[22] Consequently, the alleged deficiencies go to the weight to be given the DNA evidence, not its admissibility.

However, not all courts are as willing to admit new DNA methods relying on the precedence and proven reliability of related human DNA typing methods. For example, in October 2003, the Court of Appeals for the State of Washington ruled that animal DNA couldn't be presented as evidence in a murder trial in the same way as with human DNA evidence.[23] This court opined that the analysis of animal DNA is not as advanced or reliable as the science of human DNA. Despite relevant and interesting dicta in the final analysis, the appeals court held that the trial court's error in admitting evidence of DNA analysis performed on canine blood found on one of the defendants was harmless.

The basic facts of this case were that two victims were shot to death in their home. During this incident the victim's dog, Chief, was shot and Chief's blood was transferred to clothing worn by the defendants. PE Zoogen, a division of PE Applied Biosystems, tested three articles of clothing. Dr. Halverson concluded that for all items tested at least nine of the ten nuclear DNA markers matched Chief's DNA. Dr. Halverson testified that with a 9-marker match, the probability of finding another dog with Chief's profile was 1 in 18 billion; with a 10-marker match, the probability was 1 in 3 trillion.

The court held that the methodology and technique for extracting and identifying DNA in plants and animals is relatively well established, as is the methodology and technique for determining the probability and extent of a relationship between two samples of human DNA, including matching the DNA of a sample to a specific human being. However, the study of canine DNA has not progressed to the same extent as that of human DNA and is not sufficient to permit an expert to testify to a match between a sample and a specific dog.

During review, this court felt that the polymorphic loci in canine DNA had not been sufficiently studied such that probability estimates were appropriate for use in forensic casework. Existing canine DNA data had been derived and used for paternity testing, pure breed testing, and cancer and disease studies. While the data may be sufficient for these purposes, it was not for criminal applications. What the trial court should have established but failed to do was whether identifying a dog through canine DNA with the same degree of forensic accuracy of identifying a human is accepted in theory and practice by the relevant scientific community. Specifically, the inquiry should have established the following:

- What was known about the nature and extent of genetic variation in canines
- Whether the scientific community generally accepted that the specific loci used by PE Zoogen in the present case were highly polymorphic and appropriate for forensic use
- Whether the scientific community generally accepted that PE Zoogen's calculations of a random match were based on accurate probability estimates derived from an appropriate dog database.

In the final analysis it was determined that based on the available information, the court was not convinced that forensic canine DNA identification is a theory that has received general acceptance in the scientific community, or that reliable techniques or experiments exist to identify individual canines for forensic purposes. Generally what should be inferred from this judicial ruling is not a ban on the use of DNA analysis performed on canine or any other nonhuman DNA source of evidentiary value but rather that sufficient research and validation must be completed to determine that the specific loci used meet all of the necessary forensic criteria. Studies demonstrating that the derived databases are in Hardy–Weinberg and linkage equilibrium, the markers are highly polymorphic with sufficient discrimination power, and that population databases are relevant to the evidence when making a comparison are all important issues that will continue to be challenged for newer DNA technologies.

13.8 Future Legal Issues Relating to Nonhuman DNA Testing

Many current DNA typing methodologies are extremely sensitive, requiring only a minute amount of sample to yield a genetic profile. Generally, this is a good characteristic for forensic samples; however, this low threshold also gives rise to concerns regarding contamination by exogenous DNA sources. Mitochondrial DNA-based methodologies are particularly susceptible to contamination issues. While contamination issues are significant and serious, and must always weigh heavily in the minds of crime scene personnel as well as laboratory personnel, courts have generally not viewed contamination concerns as fatal to admissibility. So long as the basic methodology is reliable and was properly applied in the case at question, then issues regarding contamination go to the weight of the evidence by the trier of fact, most often the jury.[24]

Likely we will see an increased use of and acceptance of results obtained from nonhuman DNA samples associated with criminal investigations. Moreover, we will see a greater diversity of the types of DNA-containing samples. Bacteria, pathogens, pollen, various plant-based food items (e.g., herbs), mold and fungi, and any of a greater variety of domesticated and wild animal hairs will become targeted samples. Likely these variants will arise when a specific case dictates the need for such analysis — that is, when a nonhuman DNA sample is located at a crime scene or during the investigation and it is determined that analysis may provide valuable investigative linkages or associations, or exclusions.

The extent to which the DNA analysis may yield valuable information will greatly depend on whether the selected materials have been subjected to sufficient testing and validation. If statistical match criteria are to be employed, then the unknown sample will need to be compared to a relevant statistically valid database that has been demonstrated to contain polymorphic alleles that are in Hardy–Weinberg and linkage equilibrium. Regardless of the source of nonhuman DNA, as Sensabaugh and Kaye proposed, there are certain factors that the courts will examine in judging the admissibility of novel DNA evidence based on precedents already established for human DNA identification.[25] Generally, these inquiries will focus on the underlying methodology and its application to the case at hand, and in the validity of the statistical interpretations. As with many cases involving human DNA, once these foundational aspects are established, challenges will be made on a case-by-case basis raising issues of contamination, or in the reliability of the laboratory and/or scientist who performed the actual testing.

References

1. Wambaugh, J., *The Blooding*, Perigord Press, New York, 1989.
2. Cited within 536 Pa. 508, 519. U.S. Department of Justice, Office of Justice Programs, Bureau of Justice Statistics, Forensic DNA Analysis: Issue (Government Printing Office, Washington, D.C., 1990).
3. *Frye v. United States*, 293 F. 1013 (D.C. Cir. 1923).
4. *People of the State of New York v. Joseph Castro*, 545 N.Y.S.2d 985.
5. *Frye v. United States*, 293 F. 1013 (CA DC 1923).
6. Bohan, T.L. and Heels, E.J., The case against Daubert: the new scientific evidence "standard" and the standards of the several states, *J. Forensic Sci.*, 40, 6, 1995. These authors cited *Ex Parte Perry*, 586 So.2d 242 (AL 1991) as a case that explicitly identified the three prongs to a Frye analysis.
7. See Supra note 6, 1031.
8. *United States v. Downing*, 753 F.2d 1224, 1237 (1985).
9. Meaney, J., From Frye to Daubert. Is a pattern unfolding?, *Jurimetrics J.*, 191, 194, 1995.
10. *Daubert v. Merrell Dow Pharmaceuticals, Inc.*, 113 S.Ct. 2786; 61 U.S.L.W. 4805 (1993).
11. *State of Connecticut v. Christian E. Porter*, 241 Conn. 57 (1997); 1997 Conn. Lexis 155.
12. *General Electric Co. v. Joiner*, 522 U.S. 136, 138–39, 143.
13. *Kumho Tire Co. v. Carmichael*, 526 U.S. 137 (1999).
14. Starrs, J.E., Out with the old — in with the new, will the evidence rules revisions outmuscle Daubert-Kumho?, *Scientific Sleuthing Review*, 24, 2000.
15. Federal Rules of Evidence 403 and corresponding state rules of evidence state that relevant evidence may be excluded if its probative value is outweighed by the danger of unfair prejudice or surprise, confusion of the issues, or misleading the jury, or by considerations of undue delay, waste of time, or needless presentation of cumulative evidence.

16. *State of Connecticut v. Pappas*, 256 Conn. 856; 776 A.2d 1091: 2001 Conn. Lexis 284.
17. *People of the State of Michigan v. Kevin Holtzer*, 255 Mich. App. 478; 660 N.W.2d 405; 2003 Mich. App. Lexis 523.
18. *State of Arizona v. Bogan*, 183 Ariz. 506; 905 P.2d 515.
19. Savolainen, P. et al., Sequence analysis of domestic dog mitochondrial DNA for forensic use, *J. Forensic Sci.*, 42, 593, 1997.
20. Menotti-Raymond, M. et al., Genetic individualization of domestic cats using feline, STR loci for forensic applications, *J. Forensic Sci.*, 42, 1039, 1997.
21. *United States of America v. Boswell*, 270 F. 3d 1200; 2001 U.S. App. Lexis 2235.
22. *United States v. Beasley*, 102 F.3d 1440, 1448 8th Cir. 1996.
23. *State of Washington v. Kenneth Leuluaialii, a/k/a/George Tuilefano*; 77 P.3d 1192; 2003 Wash. App. Lexis 2335.
24. *The People of the State of New York v. Edmund Ko*, 304 A.D. 2d 451, 2003 N.Y. App. Div. Lexis 4201. Also see *People v. Klinger*, 185 Misc.2d 574, 713 N.Y.S.2d 823.
25. Sensabaugh, G. and Kaye, D.H., Non-Human DNA Evidence, *Jurimetrics J.*, 38, 1, 1998.

Forensic Palynology

LYNNE A. MILNE, VAUGHN M. BRYANT JR., AND
DALLAS C. MILDENHALL

Contents

0-8493-1529-8/05/$0.00+$1.50

14.1 Introduction

Palynology, the study of pollen, spores, and other acid-resistant microscopic plant bodies collectively known as *palynomorphs*, is an interdisciplinary field with applications in many areas of science, including forensics, geology, geography, botany, zoology, archaeology, and immunology. Many of these areas impact on and overlap with each other and, when combined, form the study of forensic palynology. Like many forensic disciplines today, palynology involves the study of microscopic evidence that is resistant to damage or removal from crime scenes. The value of forensic palynology lies in four important attributes of pollen and spores:

- They are microscopic in size.
- Pollen and spores are produced in vast numbers.
- Pollen and spores can be identified to a plant taxon.
- They are highly resistant to decay.

Therefore, someone who commits a crime may unknowingly take away or leave at a crime scene critical pollen and spore evidence in large numbers and they are too small to be seen. Pollen and spores are so resistant to decay that they will be preserved for many years without specialized storage, and so morphologically complex that they can be identified to a specific plant type, site, region, or country (Figures 14.1 and 14.2; see color insert following page 238).

All plants produce either pollen or spores as part of their reproductive cycle. Pollen, the fine yellow dust produced by flowers, is best known for causing hay fever and contributing to asthma. These misery-producing microscopic particles are the plant world's equivalent of sperm — they carry the male sex cells of the higher vascular plants (flowering and cone-bearing plants). Spores, although similar in size and distribution to pollen, are the asexual reproductive bodies of lower vascular plants (e.g., ferns) and non-vascular plants (e.g., mosses and fungi), and develop into a new organism without the need for joining with another reproductive cell.

To effect fertilization, pollen grains generally reach the female parts of a flower of the same species with the assistance of wind or animal vectors (typically insects). As this is a random affair, plants must produce enormous amounts of pollen to ensure that some pollen grains reach their intended destination (the receptive stigma of a flower). Most pollen grains are unlucky and end up as components of soil and dust, falling onto any surface exposed to the air. The type of pollen present at any one location, or on any object, can often be related to a particular vegetation type, site, or region. It is this aspect that enables the comparison of pollen samples to prove or disprove a relationship between a suspect, victim, and people (such as witnesses), objects, or localities.

14.2 History and Utilization of Forensic Palynology

One of the first successful uses of forensic palynology that gained widespread public attention occurred in Austria in 1959.[1] A man disappeared on a trip down the Danube River and police suspected foul play. The pollen assemblage retrieved from mud on a suspect's boots indicated a particular forest environment based on the composition of pollen types from the boot sample. Particularly informative, the assemblage also contained a single fossil pollen grain from the Miocene age. Rocks in which this fossil pollen type was known

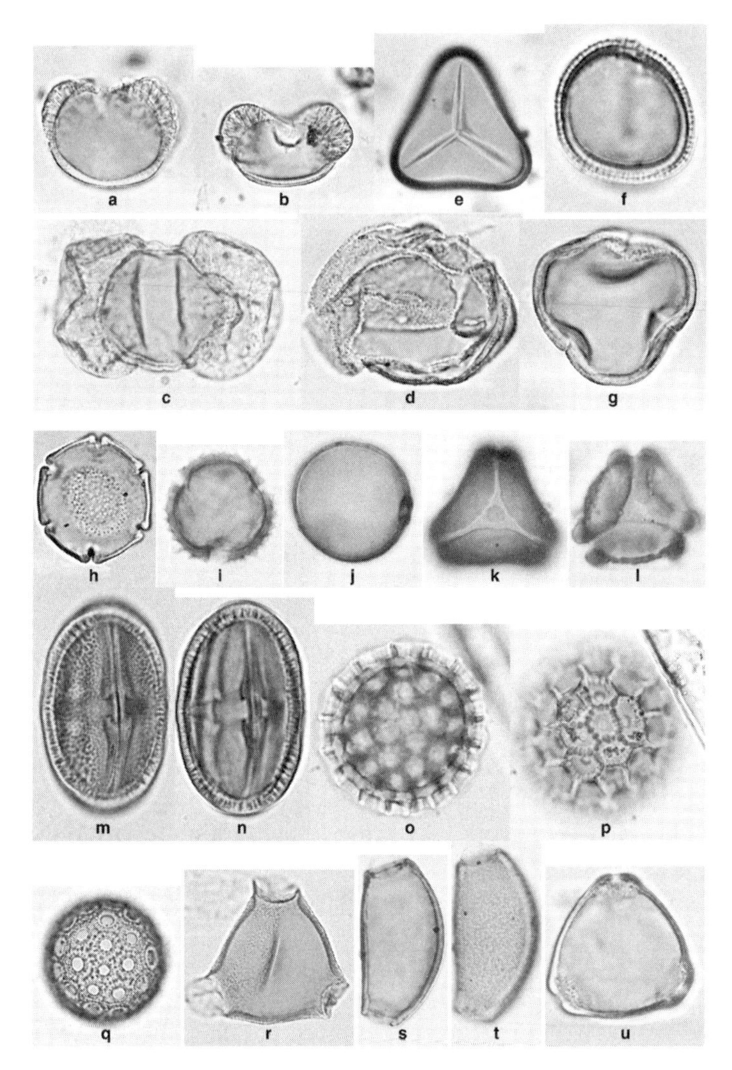

Figure 14.1 Transmitted light photomicrographs of conifer and angiosperm pollen from New Zealand and Australia, and a fern spore. Scale bars = 10 μm. Magnifications range from 600X to 900X. a–d, Conifers; a. *Dacridium*; b. *Halocarpus*; c. *Podocarpus*; d. *Agathis*, e. Fern spore; Cyathea. f–u, Angiosperms. f. *Ascarina* (Chloranthaceae); g. *Coprosma* (Rubiaceae); h. *Nothofagus* (Fagaceae); i. *Olearia* (Asteraceae — "daisy"); j. grass (Gramineae); k, l. Myrtaceae — e.g., *Eucalpytus*; m, n. *Scaevola* (Goodeniaceae), same grain, high and median foci respectively; o, p. *Tribulus* (Zygophylaceae), same grain, median and high foci respectively; q. *Ptilotus* (Amaranthaceae), high focus; s, t. *Banksia* (Proteaceae), same grain, median and high foci respectively; u. *Casuarina* (Casuarinaceae). Photography: a, b, d–h, D. Mildenhall; c, i–u, L. Milne.

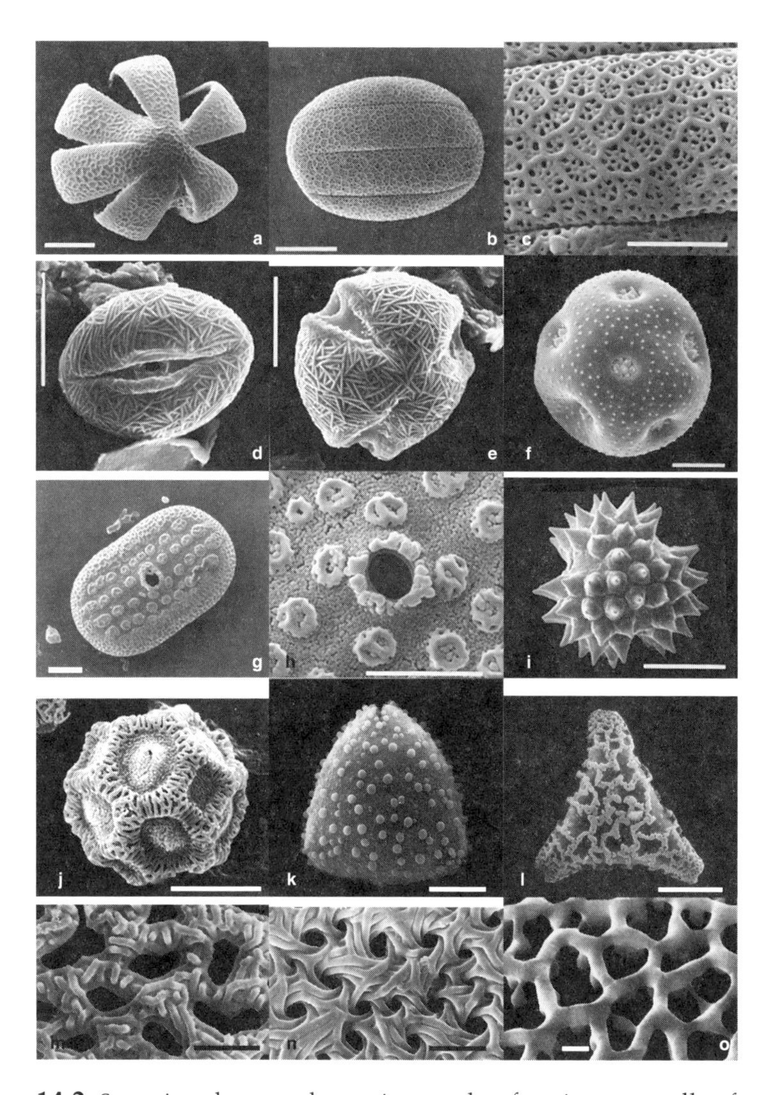

Figure 14.2 Scanning electron photomicrographs of angiosperm pollen from the southeastern U.S. and Australia. Scale bars: a–i = 10 µm, l = 50 µm; m–o = 1 µm. Magnifications range from 700X to 11,500X. a. *Lamium amplexicaule* (Lamiaceae); b, c. *Salvia azurea* (Lamiaceae); b. whole grain; c. surface ornamentation; d, e. *Sedum nuttallianum* (Crassulaceae), equatorial and polar views respectively; f. *Minuartia drummondii* (Caryophyllaceae); g, h. *Justicia runyonii* (Acanthaceae); g. whole grain, h. surface ornamentation; i. *Solidago canadensis* (Asteraceae); j. *Opuntia stricta* (Cactaceae); k. *Beauprea montis-fontium* (Proteaceae); l. *Proteacidites reticularis* (fossil Proteaceae); m–o, Surface ornamentation of three species from the same Proteaceae genus; m. *Petrophile media*, n. *Petrophile conifera*, o. *Petrophile diversifolia*. Photography: a–j, adapted from Jones, G.D., Bryant, V.M., Jr., Hoag Lieux, M., Jones, D.J., and Lingren P.D., *Pollen of the Southeastern United States,* American Association of Stratigraphic Palynologists Foundation, Contributions Series, 30, AASP Foundation, 1995; k–o, L. Milne.

to occur were associated with only one region of the forest located in a small area along the Danube River. When confronted with this pollen evidence, the suspect confessed and led police to the gravesite.

The forensic applications of palynology are almost limitless. It has been utilized in such cases as homicide, rape, assault, forgery, armed robbery, drug dealing, and fraud. However, because it is a specialty area of forensic science requiring unique training, it is still a much underutilized forensic tool. In New Zealand, forensic palynology has been conducted routinely for almost 25 years, and it is increasingly becoming an accepted part of forensic science in the U.S., U.K., Australia, and other countries. Although pollen analysis can and has helped convict criminals, its greatest application is for providing corroborative or associative evidence and for providing investigative leads. Pollen evidence also has, in many cases, prompted suspects to confess to their crimes. In this sense, palynology may provide law enforcement with simple and inexpensive evidence for case investigations.

This chapter will deal primarily with pollen, but most of the aspects discussed also apply to spores. The biological role of pollen, its morphological complexity, and production and dispersal patterns are discussed in detail. This basic knowledge will assist crime scene personnel in critically assessing the value of palynology for a particular case. Also outlined in this chapter are the mechanics of pollen preparation and analysis, and its strengths and limitations for forensic analyses. For the crime scene investigator, sample type and sample collection and storage will be discussed.

14.3 Pollen and Spores — What, Where, and Why?

14.3.1 The Biological Role of Pollen

The male parts of a flower, the stamens, consist of a stalk (the filament) with a small sac (the anther) at the tip. Pollen is produced in the anthers, and each grain contains two nuclei. The female reproductive organ of a flower is the pistil, which consists of the ovary, stigma, and style. The ovary contains one or more ovules, each of which contains a female sex cell. The stigma is the receptive organ and has a sticky surface that captures pollen. The style connects the stigma to the ovary. When a pollen grain lands on the stigma of a flower of the same species, one of its nuclei forms a pollen tube through the tissue of the style and into the ovary. The other nucleus divides in half to form two male nuclei, which then travel down the pollen tube to an ovule. One of the male nuclei fuses with the female cell in the ovule (fertilization), and the other fuses with additional nuclei in the ovule to form the endosperm that provides the developing plant's food store in the seed.

14.3.2 The Physical Characteristics of Pollen

To the naked eye, pollen appears to be a homogenous mass of fine yellow dust. When examined with a microscope, it becomes apparent that this mass actually consists of thousands of individual pollen grains with a shape and form that is often unique to the plant species that produced them. Although microscopic in size, pollen grains come in a vast array of shapes and sizes, and many have exceptionally complex surface patterns. It is this wide variety of physical features that enables palynologists to identify pollen and spores as coming from a unique plant family, genus, and often species. Because of their complex morphology, palynologists have sometimes called pollen and spores "the fingerprints of plants."

For the routine examination of pollen samples, transmitted light microscopy (TLM) with magnifications of up to one thousand times the original size of the object is adequate. Scanning electron microscopy (SEM) is used when higher magnifications of fine surface details are needed to distinguish the subtle, minute differences between pollen grains of different plant species. The physical characteristics of pollen that are used by palynologists to differentiate the pollen of one plant taxon from that of another are discussed in the sections that follow.

14.3.2.1 Size

Pollen and spores range in size from 5 to 200 μm in diameter, but most fall within the 20 to 70 μm size range. There are 1000 μm (micrometers or microns) in 1 mm (millimeter). For many investigators who do not use microscopic analysis on a regular basis, it may be difficult to comprehend something so small yet so complex. To put the size of pollen in perspective, imagine carefully placing pollen grains end to end in a straight line to make a line 1 mm in length. It would take one hundred pollen grains of 10 μm each in diameter to make that line. As Wodehouse[2] so aptly put it, "Pollen grains belong to a world of another size."

The pollen grain diamcter of each plant species generally has a fairly limited size range. At the plant family level, size range may be quite large, but further down the taxonomic hierarchy the size range decreases. For example, the pollen of a genus may have a size range difference of nearly 20 to 50 μm, while a species within the genus may have a smaller size range of only a few micrometers. The overall size of a pollen grain from a species may help to differentiate it from other species within the same genus. Size is a feature that helps to place a pollen type in a general category (genus), yet the potential for size variations within a genus makes size by itself an unreliable way to identify pollen and spores to the plant species level.

14.3.2.2 Shape

Pollen grains come in many shapes. They may be spherical, triangular, elliptical, hexagonal, or pentagonal or exhibit any number of variations on basic geometrical themes. For example, a triangular pollen grain may be convexly triangular, concavely triangular, or triangular with protrusions from the apices or sides. Beyond the general shape, there are variations in the ratio of the polar and equatorial diameters, such that some grains are thinner in one plane than the other. Like size, shape is useful for helping to place a pollen type in a general category, but shape alone is not usually sufficient for a precise identification to the plant species level.

14.3.2.3 Aperture Type

Most pollen grains have one or more small openings (apertures) for the male gametes to escape through during the process of pollination. A few pollen types have no apertures and instead have thin walls that rupture to allow the male gametes to escape. Apertures range from a single circular hole (pore) or slit (colpus) to numerous and complex combinations of pores and colpi. In general, the aperture type is broadly consistent within many plant families, but more so at the genus and species levels. Finer structural details of apertures such as size, number, shape, and the thickness and stratification of the surrounding walls are major distinguishing features between different genera and between different species from the same genus.

14.3.2.4 Sculpture

Sculpture refers to the ornamentation or pattern on the surface of pollen grains. It may be quite simple or very ornate, but it is the same for pollen from plants of the same species. Pattern types consist of variations in the depressions on the surface, or protrusions from it. For example, some of the protrusive forms of patterning are termed granulate (small, sandlike protrusions), echinate (spines), verrucate (hemispherical wart-shaped bodies), baculate (rodlikc), or reticulate (a lacelike network). Some pollen grains, such as those from grasses and sedges, have surfaces that appear almost smooth when viewed with a light microscope, but distinct surface patterns emerge when they are examined with electron microscopy under increased magnification and resolution.

14.3.2.5 Wall Composition and Preservation

Pollen and spore walls are largely composed of cellulose, protein, and various amounts of a tough biopolymer called sporopollenin, which is resistant to most acids but is susceptible to oxidation. Pollen grains and spores are best preserved in an acidic, anaerobic environment, but they will gradually deteriorate in aerobic, alkaline environments. The protoplasm inside pollen and spores decomposes quickly, but the outer wall may remain preserved for many millions of years if the grains are buried rapidly in an anaerobic, acidic environment. Even in adverse depositional conditions, such as oxidizing environments that are cyclically wet and dry, most pollen will survive in surface soil and leaf litter for at least a few years. It follows that plants do not need to be flowering at the time a crime was committed for its pollen to be found in soil, on the plant itself, or on articles found at or removed from a crime scene. This resistance to degradation also means that forensic evidence that is collected and stored correctly for many years can still retain pollen that can link the evidence back to a crime scene.

14.3.2.6 Wall Structure

The pollen wall, termed the exine, is stratified and may have up to three layers, which have different chemical compositions. The two inner layers, termed the intine and endexine, may or may not be present in particular plant species, and are best seen in pollen that has not been chemically prepared for pollen analysis. The intine, the innermost layer, is made of cellulose like the pollen walls of many water-pollinated plants, and does not survive the chemical procedures used prior to pollen analysis. The next layer, the endexine, may be internally laminated or patterned, and if present in a particular species, it may occasionally be evident using light microscopy. To study these inner wall layers in any detail, it is necessary to use transmission electron microscopy. The outermost layer of a pollen wall, the ectexine, is the layer that contains sporopollenin. It is this layer that can be observed when processed pollen is examined using light microscopy.

Although there are many variations on the general theme, the ectexine typically consists of three layers arranged much like a colonnade. The foot layer forms the base from which a layer of columns or grains, called the infratectum, project to support a relatively solid or perforated top layer called the tectum. This outermost layer is quite variable and may either form the actual surface sculpture of a pollen grain, like that of reticulate pollen, or support supratectal sculptural elements if it is relatively solid. In rare instances, the tectum may not be present at all. The variations in the width and internal structure of each of these layers, relative to each other, are important diagnostic features that can be used to distinguish differences in the pollen of two very closely related species.

14.3.3 Pollen Production and Dispersal

Pollen has various modes for transport or dispersal, and which mode employed dictates how much pollen a plant usually produces. Knowledge of the mode and the average amount of pollen produced are important in forensic palynology. If the pollen production and dispersal patterns of plants in a particular area are known, then one can predict what to expect in that

Table 14.1 Summary of Pollen Dispersal Methods, Expected Levels of Pollen Production, and Their Forensic Potential

Method	Grains per Anther	Occurrence in Forensic Samples	Forensic Potential
Water (Hydrogamous)	1000s	Lack sporopollenin, rarely preserved	Little to none unless collected as part of a water sample
Self (Autogamous)	<100	Rare in sediments near plants	Rare, but very specific if present
Closed (Cleistogamous)	<100	Very rare near plants	Rare, but excellent if present
Animals (Zoogamous)	100–1000	Rare to common near plants	Excellent
Wind (Anemophilous)	1000–>100,000	Common near plants and in neighboring areas	Excellent to poor, depending on the plant associations

area's pollen assemblage. If unexpected pollen and spore types are present in unusual numbers, research may be needed to examine why atypical species are present. A basic understanding of pollen production and the dispersal patterns of major plant types is exceptionally useful to crime scene personnel when deciding if pollen analysis will be of probative value at a given crime scene. The methods of pollen dispersal, their relative pollen production levels, and their potential for preservation become essential criteria for determining the forensic potential of pollen from different plants (see Table 14.1).

14.3.3.1 Water-Pollination (Hydrogamous)

Many submerged water plants rely on water currents to disperse pollen grains. This is an unreliable method for fertilization, so as a result, anthers produce thousands of grains to compensate for the extreme numbers of pollen grains that never contact a receptive stigma. Pollen is generally small and unornamented, with thin walls composed of cellulose. This type of pollen is of little forensic value as it does not preserve well, and does not typically survive the chemical preparation techniques used to prepare pollen for analysis. If a crime scene or a victim is associated with a lake or stream where submerged plants might be pollinating, then these types of pollen grains might provide useful evidence. If their presence is suspected in a sample, then the sample should be microscopically examined prior to chemical preparation, and processed using acid-free techniques.

14.3.3.2 Self-Pollination (Autogamous)

In self-pollinating plants, the male and female parts (the anthers and stigma) mature at the same time. The flowers either have an adaptive mechanism that will affect the pollination process itself, or they rely on wind or animals to move the pollen the short distance from the anthers to the stigma. This is an efficient method of dispersal, so only one hundred or less pollen grains are produced per anther. Pollen from these plants might be found in soil near the plant, but because only a few grains are produced they will be rare by comparison with pollen from wind- and animal-pollinated plants.

14.3.3.3 Closed-Pollination (Cleistogamous)

This type of pollination occurs within flowers before they open, or in flowers that do not open at all. The pollen is often large with very little surface ornamentation, and less than one hundred grains are produced per anther. Many cereal plants are cleistogamous, but in some types of wheat a few anthers often emerge after the seeds have set. Why this happens is a mystery, but for forensic science, and in particular for archaeological studies, it is a bonus, since a couple of cereal pollen grains in a soil sample may add a lot of information.

14.3.3.4 Animal/Insect Pollination (Zoogamous)

The stigma and anthers of zoogamous plants tend to mature at different times to guard against self-pollination. Their goal is to pollinate a flower of a different plant of the same species, and to this end they produce relatively ornate flowers and conspicuous amounts of nectar to attract their pollinators. Bees, ants, beetles, butterflies, wasps, bats, lizards, and small terrestrial mammals may unwittingly transfer pollen from one flower to another during their search for food and nectar. This is a relatively efficient method of pollen transfer, so plants produce only a few pollen grains, usually approximately one thousand or less per anther. These types of pollen usually have thick walls, are relatively heavy and often highly ornamented, and typically preserve well in soil. More than half the flowering plants in the world use this method for pollination. The forensic potential of pollen from zoogamous plants is excellent because the pollen only occurs in small amounts near or on the plants that produced them (few if any are carried by wind currents), and their highly diverse designs enable many to be identified to the genus or species level. When these types of pollen are found on clothing, it likely means that it was deposited when a person brushed against the plant or came in direct contact with soil near or under the plant.

14.3.3.5 Wind-Pollination (Anemophilous)

This is a very inefficient method of pollination, and plants must produce vast quantities of their pollen to ensure that a few of the pollen grains reach their intended destination. Many wind-pollinated plants such as grasses, pine, oaks, birch, alder, elms, and some *Eucalyptus* species produce between 10,000 and >100,000 grains per anther.[3] The female *Cannabis* plant produces about

70,000 pollen grains per anther,[4] and a whole plant can produce more than 200 million pollen grains. Wind-pollinated plants generally release their pollen grains early in the morning once the sun's heat dries the anthers and the heated ground creates wind currents and updrafts. The heaviest concentration of pollen in the air usually occurs during periods of low humidity in the late morning when the pollen is released, and again near sunset when the ground cools and reduced wind allows the dispersed pollen to sink back to the ground level. Wind-pollinated plants include conifers and some flowering plants, in particular many forest trees, sedges, and grasses. Ferns, mosses, and fungi, the spore-producing plants, similarly produce vast quantities of spores and use wind to disperse them. Because of the large volume of dispersed pollen (and spores), wind-transported pollen is usually deposited on almost every exposed surface and is the most likely type to be deposited on the ground and become part of the permanent pollen deposits for a record of a geographic region.

The ability of a pollen grain to remain aloft and travel on air currents is determined by its "sinking rate." The size of a pollen grain, its aerodynamic shape, overall mass, and the height at which it is dispersed will determine its sinking rate and how far it will potentially travel from the parent plant. Other factors that will influence the distance pollen may travel from its source include wind currents, humidity, and obstacles in the path of the pollen grains. Studies show that for many species, 95% of their wind-transported pollen falls between 25 m and 2 km from the parent plant, and the remaining 5% usually falls between 2 to 100 km away from the source.[5] Sometimes a few pollen grains may reach the upper levels of the atmosphere where they can be carried thousands of kilometers. The forensic potential of wind-transported pollen ranges from excellent to poor, and depends on the uniqueness of the association of plant types in a particular geographic area.

14.4 Pollen Analysis — How It Works

14.4.1 Palynomorph Assemblages: The "Fingerprints" of Localities

A palynology sample from anything but a single plant species will contain much pollen and many spore types, as well as a range of other microscopic plant material. The total number of pollen and spores identified in a sample is called a "pollen assemblage," or palynomorph assemblage. Samples from a terrestrial environment can contain pollen, fern and moss spores, fungal spores, plant cuticles, wood fragments, fibers, tracheids, phytoliths (silica crystals produced by most plants), possibly fragments of charcoal, and even insect remains. All of these microscopic particles may be of significance for the comparison of forensic pollen assemblages.

Samples that have been in or near open water sources often contain pollen and spores characteristic of that source. Even tap water may contain pollen and spores. Water samples may also contain other indicators that suggest a locality or particular environment. Some of these non-pollen and spore indicators may include cysts from microorganisms called dinoflagellates, as well as diatoms (microscopic unicellular algae that produce a siliceous skeleton). In marine environments, the calcareous remains of microscopic protozoans (foraminifera) may be found in samples that have not been chemically treated with acids. The chitinous organic lining of these small animals does survive chemical pollen preparation, and if present in a processed sample, they are excellent environmental indicators.

14.4.2 Sample Preparation for Pollen Analysis

The extraction of pollen and other palynomorphs from sediment (soil, mud, sedimentary rock) and samples taken from forensic evidence involves sequential chemical procedures. The objective is to remove as much mineral material and large extraneous organic matter as possible from the sample, leaving only the finer organic matter, which will contain the pollen and spores. The amount of sediment used depends on its type (e.g., clay, sand, or limestone) and the case in question. Even a small pinch of sandy soil or clay may contain pollen, so often it is not necessary to process large amounts. Sometimes a few grams of material are all that is needed for processing, but in other cases slightly larger samples might need to be processed.

The order in which the chemical procedures are conducted on a sample depends on the historical protocol of each laboratory, the type of materials being examined, and the desired results. In summary, calcium carbonate and other carbonate minerals can be removed by adding hydrochloric acid, and silicate minerals (e.g., quartz sand, clay) are removed using hydrofluoric acid. After these procedures have been performed, all that should be left are heavy minerals and organic matter. To remove fine organic material and the protoplasm and outer lipids from pollen and spores so that their features can be clearly seen using light microscopy, the sample is taken through a procedure known as acetolysis. This involves heating the sample for three to ten minutes in a mixture of sulfuric acid and acetic anhydride in a 1:9 ratio. Because this solution is anhydrous (repels water), the sample must be rinsed several times with acetic acid before and after it is acetolysed.

Large fragments of plant material and other unwanted debris are removed from the sample residue by sieving it through a 150 to 200 μm mesh sieve. Sieving can be conducted at any stage of the preparation procedures, but it is best conducted after acetolysis to ensure that pollen grains and spores adherent to large organic fragments are not lost. Heavy minerals that have a higher specific gravity than pollen and spores may be removed from the sample by heavy liquid separation, using, for example, zinc bromide with a specific gravity of 1.65 to 2.0. During each of these steps, the sample residue is concentrated by centrifugation and the supernatant (excess liquid) is discarded. Between each chemical procedure, the sample is rinsed several times with distilled water and centrifuged each time as described earlier. The organic residue recovered from samples after chemical preparations are completed may be no larger than a match head, but it may contain thousands of palynomorphs.

The procedures described in this chapter are standard for soil and sediment samples. For forensic samples, only some of the procedures may be necessary, and a number of other chemical and laboratory procedures may be needed to free pollen from specific types of matrices. For example, some evidentiary soil samples may require deflocculation with various compounds including detergents, or may need sonication to free pollen from some types of matrix materials. Occasionally, careful treatments with various types of oxidants (e.g., nitric acid or strong bases such as potassium hydroxide) might also be necessary.

Once the extraction procedure is completed, small drops of the organic residue are permanently mounted on glass microscope slides for examination with light microscopy. Mounting compounds vary depending on a palynologist's preference and the type of analysis that is needed. Some palynologists use fluids such as glycerin or silicon oil that remain viscous and allow individual pollen and spore grains to be rolled over to search for unique morphological features. Other examiners use a permanent plastic-mounting medium so that grains can be relocated with ease on a microscope slide.

The pollen assemblage of a sample is characterized by identifying the pollen types present to the lowest possible taxonomic level (i.e., family, genus, species). To do this, the investigator compares pollen and spores in the assemblage with other types in pollen reference collections. Reference collections may be drawings and photographs in journal articles or books and from the pollen of plants that were collected at a crime scene. In addition, a growing number of Internet websites now contain images of pollen and spore types and reference keys for certain groups of plants. Once each specific pollen or spore type is identified in a sample, the relative percentage of occurrence of each type is calculated by counting two hundred or more grains. Statistically, a count of two hundred pollen grains may suffice. When searching for a rare pollen type, it might be necessary to count up to 1000 pollen grains or more per sample, or, alternatively, slides can be scanned for rarer pollen and spore types after the completion of the 200-grain count. If necessary, portions of a sample can be examined using a scanning electron microscope.

14.4.3 Interpreting Forensic Pollen Assemblages

The pollen assemblage found in a soil, dirt, dust, or mud sample from a given location will normally reflect a "partial image" of the surrounding vegetation at that site, and to some degree the wind-pollinated vegetation of the nearby region. The pollen assemblage is a "partial image" because of the different patterns of pollen production and dispersal for each plant type. Most of the pollen from a site will come from local and regional wind-pollinated plants, and the local insect-pollinated plants will be highly underrepresented or at times not represented at all. Therefore, the percentage occurrence of pollen types in a sample will not be equivalent to the percentage occurrence of their parent plants in the local vegetation.

The association of pollen types and their relative frequency in an assemblage may be unique for a given location, a region, a larger geographical area, or even a country. These pollen assemblages provide what is often called a "pollen print" for a given location. That pollen print then becomes the "control sample" for that location against which other samples, or the pollen from forensic samples, may be compared. Different types of plant communities such as forests, woodlands, and heath, and different ecological environments will produce markedly different pollen assemblages. Similarly, pollen assemblages from man-made or disturbed environments, such as pine plantations, vacant lots, and roadside verges, and different towns, suburbs, and even different suburban gardens, can be distinguished and thus provide useful evidence or investigative leads. Many pollen samples taken from forensic evidence, particularly from shoes or vehicles, contain pollen from more than one locality. By knowing the expected type of pollen assemblages from different plant communities and different environments, it is possible to determine the travel history of some items. It is also possible to recognize through pollen analysis that a secondary crime scene may not be the location where the crime was actually committed (primary crime scene). For example, if a body has been moved to another location after the crime, it may have pollen assemblages from both the primary and secondary scene associated with it.

14.4.3.1 An Australian Example

The flowers of *Banksia*, a member of the Southern Hemisphere Proteaceae family, are small but densely borne on floral spikes that range in size from a few centimeters to 40 centimeters in length. Up to 6000 individual flowers have been counted on one spike of *Banksia grandis*,[6] one of the larger species. In a *Banksia* woodland, *Banksia* are the most conspicuous plants. Also present in this vegetation type there may be gum trees (Myrtaceae), sheoaks (*Casuarina* and *Allocasuarina*), wattles (*Acacia*), grasses (Poaceae), members of the daisy family (Asteraceae), and other minor plant types. If the *Banksia* plants

Table 14.2 Comparison of Pollen Assemblages from Two Different Environments: *Banksia* Woodland and Salty Wetlands

		Banksia Woodland % Occurrence		Salty Wetlands % Occurrence	
Pollen and Spores		Regional	Scene	Regional	Scene
1.	*Casuarina*	36	34.4	12	4.4
2.	Myrtaceae	34	33.6	26	27.6
3.	Pine	0.4	1.6	13.2	4.4
4.	Grasses	7	17.2	16.4	15.6
5.	*Banksia* sp. 1	2.2	2	—	—
6.	*Banksia* sp. 2	1.8	0.4	—	—
7.	*Stirlingia*	0.4	0.4	—	—
8.	*Acacia*	1	—	—	—
9.	Chenopodiaceae	—	—	14.4	37.6
10.	*Avicennia marina*	—	—	1.2	2.4
11.	*Lepidospermum gladiatum*	—	—	2.8	—
12.	*Olearia axillaris*	—	—	4	2.4
13.	*Frankenia pauciflora*	—	—	0.4	—
14.	Spore 1	—	—	0.8	0.8
15.	Spore 2	—	—	0.4	0.4
16.	Spore 3	—	—	1.2	0.8
17.	Other — gen. et sp. indet.	17.2	10.4	2.8	3.6

were flowering, most people would expect that a pollen sample from this environment should be dominated by *Banksia* pollen. However, *Banksia* are insect-pollinated and unless the soil sample collected includes a dead insect or has *Banksia* flowers in it, a palynologist will expect less than 5% *Banksia* pollen to occur in the sample. If *Banksia* pollen is present at all, it is significant. This example illustrates the importance of understanding the mode of pollination for each plant species in order to determine if a small percentage of pollen may be significant to the sample. Table 14.2 shows how pollen assemblages can be interpreted from the pollen types present in a sample and their percentage of occurrence. The table compares two pollen assemblages, each from the same environment, with two pollen assemblages from a different environment. The percentage occurrence of the pollen types in each sample was calculated from five hundred grains.

Both of the samples from the same environment contain the same pollen types, which is to be expected from samples that have come from the same type of plant community. The percentage occurrence of most pollen types in the two samples is similar, but there are some exceptions. In the *Banksia* woodland, the percentage occurrence of grass pollen is quite different in each sample, and in the salty wetlands, the occurrence of pollen from Chenopodiaceae (the saltbush family) is dissimilar in the two samples. Both these plant types are wind-pollinated, and the differences in their occurrence in the same environment suggest that the samples may not have been taken from exactly the same location but that they originate from the same type of plant community.

The only pollen types the *Banksia* woodland and salty wetland environments have in common (nos.1 to 4) are from wind-pollinated plants that are ubiquitous in the regional vegetation of the southern region of Western Australia. Although these pollen types comprise 77 to 90% of the *Banksia* woodland pollen assemblage, and 52 to 68% of the salty wetland pollen assemblage, they are not generally indicative of a particular environment. This pollen will come from both the regional vegetation and from plants in or near the locality where the sample originates. Because wind-pollinated plants are large pollen producers, only a few of these plant types need to be present near a sample site for their pollen to predominate in a pollen assemblage. The remaining pollen types listed, with the exception of the unidentified group, occur in either one environment or the other and are present in much smaller numbers than the wind-pollinated pollen types. Therefore, these pollen types are considered environmental indicators.

In the *Banksia* woodland pollen assemblage, the presence of *Banksia* pollen is indicative of a sample taken from a site close to *Banksia* plants because *Banksia* is insect-pollinated. Even though this pollen occurs in low numbers, it suggests that the plant community may contain a number of *Banksia* species and likely a considerable number of *Banksia* plants. *Banksia* prefers a certain type of soil, as does *Stirlingia*, another important environmental indicator in the *Banksia* woodland pollen assemblage. *Stirlingia* is a member of the Proteaceae family, is insect-pollinated, and is not a garden plant, but it occurs naturally in certain types of bush land and some disturbed environments, including planted pine forests.

From the pollen assemblages listed in Table 14.2, it cannot be assumed that the two pollen assemblages from the *Banksia* woodland environment are from the exact same site. However, if they were forensically relevant samples and suspects asserted that they had not been in that type of environment, it would be enough evidence to suggest that they were lying. To determine if the pollen assemblages are from samples taken from sites within the same *Banksia* woodland, the unidentified pollen types must be identified to clarify if any of the species are specific to that particular *Banksia* woodland.

In the salty wetlands pollen assemblages, the Chenopodiaceae and the variety of species within this group (not listed) are definitive indicators that the samples came from a salty wetland area. Of particular interest in these pollen assemblages is the presence of *Avicennia marina*, a mangrove common in northern Western Australia but only known in southern Western Australia at one site. As *A. marina* is site-specific in the southern part of the state, if its pollen occurs in an assemblage from a forensic sample collected from a crime committed in that part of the state, there is little doubt that the item the assemblage was taken from has been at that particular unique site. Pollen from *Lepidospermum* (Cyperaceae) and *Frankenia* (Frankeniaceae) are also indicators of a wetlands environment, and it is possible that the three spore types present in the pollen assemblages, together with the unidentified pollen types, may be able to suggest a particular location within the wetlands.

14.5 The Significance of the Evidence

The value of the evidence provided by pollen analysis is different for each case. Probative value depends on the questions being asked, the pollen types present, the pollen production and dispersal patterns of each taxon (family/genus/species), the quality of pollen assemblage matches, and how prevalent a particular type of pollen assemblage is compared to others. Little statistical work has been done on the comparison of forensic pollen assemblages. To make pollen evidence comprehensible to nonscientists (e.g., judges and juries), Horrocks and Walsh[7] proposed using the Likelihood Ratio (LR). This ratio, using Bayes' Theorem, assesses the quality of the pollen match and how common the assemblage is for a given region. The quality of the evidence is categorized and then presented to the court in a form that nonscientists can understand — i.e., the pollen evidence does not support, weakly supports, supports, strongly supports, or very strongly supports the contention under investigation.

14.6 Situations in Which Palynology Can Assist an Investigation

Palynology can be used to

- Relate a suspect to the scene of a crime or the discovery scene.
- Relate an item left at the crime scene or discovery scene to a suspect.
- Relate an item at a discovery scene to the scene of a crime.
- Disprove or prove alibis.
- Corroborate a victim's account (e.g., of locality).
- Build a profile of a suspect.
- Narrow down a list of suspects.
- Aid police to focus their search in the right direction.
- Determine the travel history of items, drugs, etc.
- Determine the geographic source of drugs, fruit, people, or various imported cargoes.

Table 14.3 Summary List of the Types of Materials that Have Been Examined for Pollen in Forensic Cases

Air filter	Dried fruit	Paint/painted wood	Shovels
Antique furniture	Feathers	Ph testing paper	Sisal
Antlers	Fur	Packing materials	Skin
Apricots	Hair	Paper money	Soil/dirt
Bread	Heroin	Radiator	Stomachs
Cannabis	Honey	Raisins	Sugar
Carpet	Hides	Resin	Tea
Cars	Intestines	Rifle butts	Tobacco
Clothes	Leather	Ropes	Wood
Cocaine	Leaves	Rubber pads	Wool
Coffee	Magic mushrooms	Rugs	Weapons
Colons	Nuts and bolts	Sacks	Wax
Condoms	Oil	Seeds	
	Opium		

14.7 Types of Samples

14.7.1 Control Samples

Control samples (reference samples) are typically samples of surface soil, mud, or water from a crime scene, a discovery scene (other than the primary site), from a site where a crime is suspected to have been committed, or from where a suspect or victim lives or works. These samples are used by the palynologist to form a baseline of data about the expected pollen assemblage at a given locality. Once the baseline data are determined, the pollen recovered from forensic samples can be compared against the control data. Plant reference material from this list is also considered important.

14.7.2 Forensic Pollen Samples

When thinking of potential materials in forensic cases to sample for pollen, remember that pollen can be recovered from almost anything that has been exposed to or in contact with air. Even commercially processed and packaged food will often contain pollen from where the produce originated and from the water used in its preparation. Dried food such as figs, prunes, raisins, and sultanas collect ambient pollen as they dry. Some examples of items forensic palynology has been conducted on appear in Table 14.3. Each item has an interesting story. For example, pollen analysis has been used in the following cases:

- Pollen analysis of feathers in a New Zealand case was used to determine if a bird (dead on arrival!) was trapped in a shipment of breadcrumbs at the point of departure or its final destination.

- Pollen found in the stomach and intestinal contents of a victim revealed what a victim ate for his last meal.
- Analysis of honey, which is essentially pollen in syrup, has been used to determine if it is authentic (as labeled) or if it has been mixed or blended with less expensive imported honey.[8]
- Pollen collected from a car radiator showed that a relatively new car driven by its owner only in a metropolitan area had traveled outside a metropolitan area and along a particular road

14.7.2.1 *Soil, Dirt, and Dust*

Pollen trapped in the dirt, soil, or dust attached to an object may link a suspect or the object with the scene of a crime. Similarly, items left at the scene of a crime may contain pollen that can link a suspect to the crime, or indicate the area in which the suspect may live and/or work. Soil samples are especially useful in cases where the scene of a crime has distinctive, or different, vegetation from the place where a vehicle usually travels or from where an object is commonly kept, or from where a suspected person works and lives. From vehicles, soil, dirt, and dust can be collected from tires, the internal carpet, seats, mud flaps, fenders, number plates, radiators, air cleaners, and the body of the car. Shoes are especially useful if they have patterned soles that still contain soil or dirt, or have stitched leather or vinyl pieces (e.g., track shoes or sneakers) where pollen can accumulate and remain in the grooves even after a normal cleaning. Even fine dust from shoes with smooth soles may yield thousands of pollen grains.

14.7.2.2 *Case History: Muddy Motorbike*

A suspect was chased up a muddy track after abandoning a motorbike that he had used in a nearby robbery, but he escaped on foot. A day later, a man went to the local police station to reclaim the motorbike, stating it had been stolen from him. Police suspected the claimant was also the thief who had abandoned the motorbike the day before. He maintained that he had never been near the muddy track in question, but on previous occasions had been up a similar track nearby. Dirt samples were collected from the suspect's boots, both tracks, and from the soil of the farm where he resided. The pollen assemblages found in the soil samples from the suspect's boots and pollen from the soil sample of the track from where the thief was chased matched, whereas the pollen spectra from the alibi track and farm soils were quite different.[9,10]

14.7.3 Clothing and Woven Material

Pollen trapped in clothing is especially useful if a person who committed a crime came in contact with soil at the crime scene or brushed against flowering plants, especially plants actively releasing pollen. Almost all kinds of clothing become good pollen traps, but those with loose fibers (e.g., wool, polyester, raw cotton) make the most efficient pollen traps. Hessian, cloth bags, and other objects made of woven fabric are also potential pollen traps. Woven clothing may even retain small amounts of pollen after several normal washes. Dry-cleaned clothing or garments made from plastic or smooth, shiny material will rarely retain much pollen. Coats that have been worn many times and cleaned infrequently are usually of little value because they often contain a mixture of pollen from many locations. This makes it difficult to link them to a crime scene unless the crime scene contains a unique plant type and some of its pollen is found on the coat.

14.7.3.1 Case History: Algal Assault

A man was physically assaulted near a river in southern Queensland, Australia. Both the victim and assailant rolled into the river, but the suspect asserted that he had not been anywhere near the river when authorities later questioned him. An analysis of the victim's and suspect's clothing revealed that both had the same assemblage of algal cysts and other palynomorphs found in control samples taken from the river area. When presented with this botanical evidence, the man confessed to committing the assault.

14.7.4 Fibers from Rope, Carpets, Fur, and Animal and Human Hair

Pollen can become trapped between strands of hair and among the fibers of many products, including carpets, draperies, rope, etc. These become excellent pollen traps if they have not been thoroughly washed or combed prior to examination. Hair and fibers will often contain a combination of ambient (airborne) pollen from the area where an object originated.

14.7.4.1 Case History: Sheepish Sale

Three hundred sheep were stolen from a New Zealand farm. One week later, a farmer who lived further north put 350 sheep up for sale. This was a suspicious act as his farm was not large enough to support that many sheep. The farmer who lost his sheep inspected the sheep for sale and believed they were originally from his farm, but they bore no identifying marks to prove his claim. Wool was sheared from the top of the backs and head region of several sheep and was sent for pollen analysis. Control samples were collected from the soils of the farm where the sheep were stolen, and from the soils of the second farm belonging to the person selling the sheep. The pollen assemblage from the wool matched the pollen in the control samples from the first farm, not the second farm where they were being held for sale.[9,10] This botanical evidence helped to support the first farmer's claim to the sheep.

14.7.5 Illicit Drugs

Ambient pollen from where the raw materials of drug items may originate, and from locations where they have been processed or packaged, is often mixed with the drugs themselves. *Cannabis* leaves are covered in tiny siliceous (cystolith) hairs that act as excellent traps for pollen from the vegetation surrounding their growing area. Pollen assemblages from *Cannabis* seizures have proven useful in determining whether separate seizures in several locations are from the same crop or shipment or represent different sources of origin, and whether a *Cannabis* seizure has been grown indoors (hydroponically) or naturally in an outdoor environment.

Cannabis produces enormous amounts of pollen, and even though most growers remove male plants from their crops, some pollen often persists in the soil and/or in the dust in a building after marijuana plants have been removed from the area. This is useful when it is suspected that a building or an area has been used for growing, drying, or packaging *Cannabis*. Interestingly, the tiny hairs from *Cannabis* leaves stick to clothing or shoes, and can provide evidence that a person was involved in the propagation or packaging of the illicit drug material. Because the hairs are primarily composed of silica, sample preparation for these types of samples should not include the use of hydrofluoric acid.

14.7.5.1 Case History: Canned by Cannabis Pollen

A family well known to the police for unusual activities regularly bought and sold farm properties around New Zealand. Between real estate sales, the family did not appear to have any visible means of economic support. It was suspected that they were involved in the hydroponic cultivation of *Cannabis*. Police raided their current property and a hydroponic *Cannabis* growing operation was discovered. Pollen analysis of dust from a farm building owned by thc family over the previous 5 years also showed evidence of *Cannabis*. Authorities were then able to seize the family's assets based on an estimate of their total income from *Cannabis* production, past and present.

Cocaine has also been found to contain pollen from the geographic region in which it was grown or packaged. If pollen types are rare or specific to a region, they can be informative for determining a general geographic point of origin. Coca plants are generally dried outdoors, so they often collect pollen from the vegetation surrounding their growing and drying area. The process of extraction for the desired alkaline compounds from the coca leaves is usually conducted outdoors where additional ambient pollen may be deposited on the processed materials. Heroin will likely also contain pollen from the area in which it was processed and packaged, especially since the poppy bulbs are scarred to extract the sticky alkaline compounds that are later refined into the heroin drug. These sticky surfaces form excellent pollen

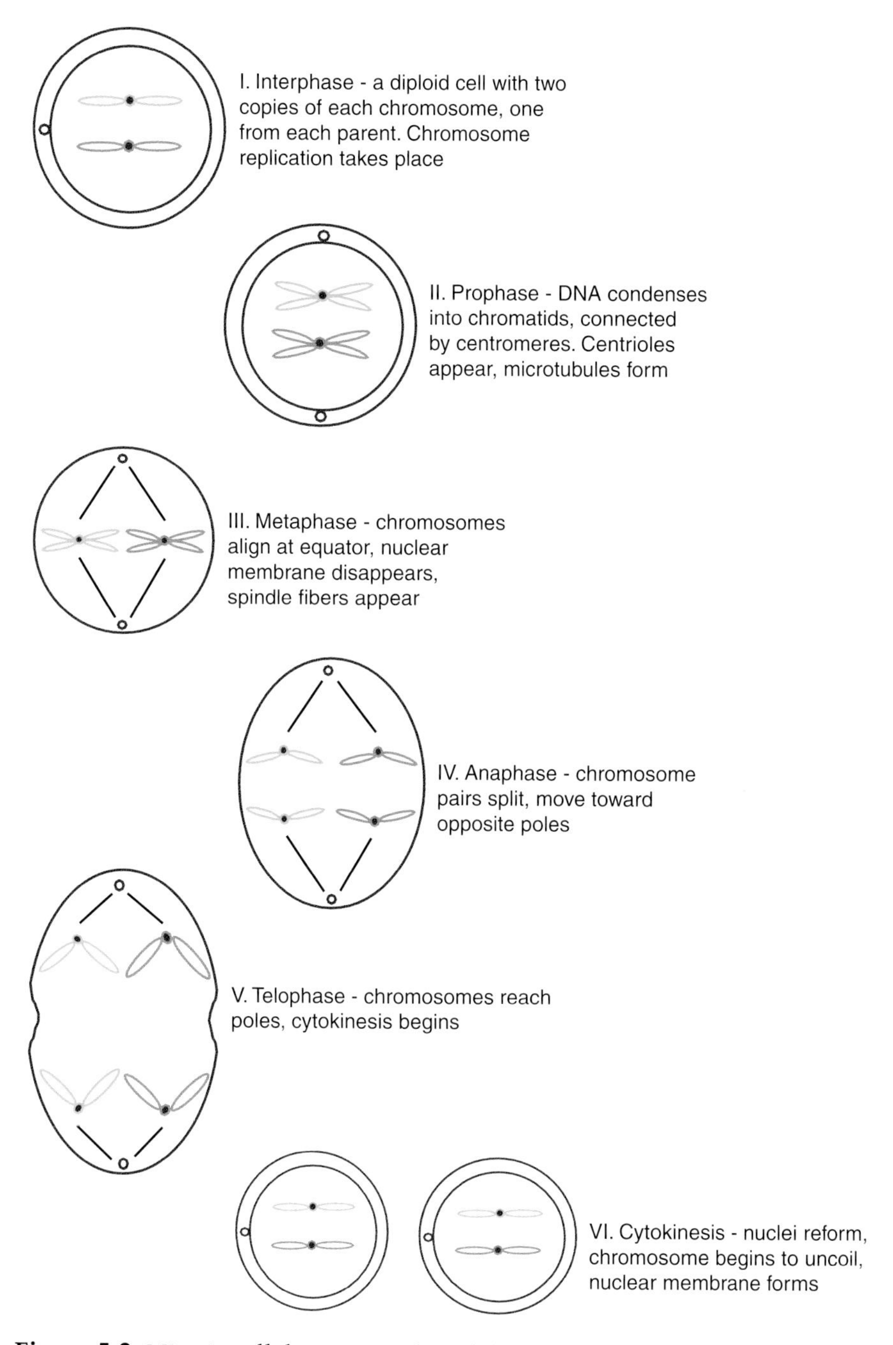

Figure 5.2 Mitotic cell division can be subdivided into different stages based on the microscopic morphology of the chromosomes: prophase, metaphase, anaphase, telophase, and cytokinesis (final cell division). Mitosis occurs in all cells that are not involved in sexual reproduction and normally results in an exact replication of DNA content in each of the two daughter cells.

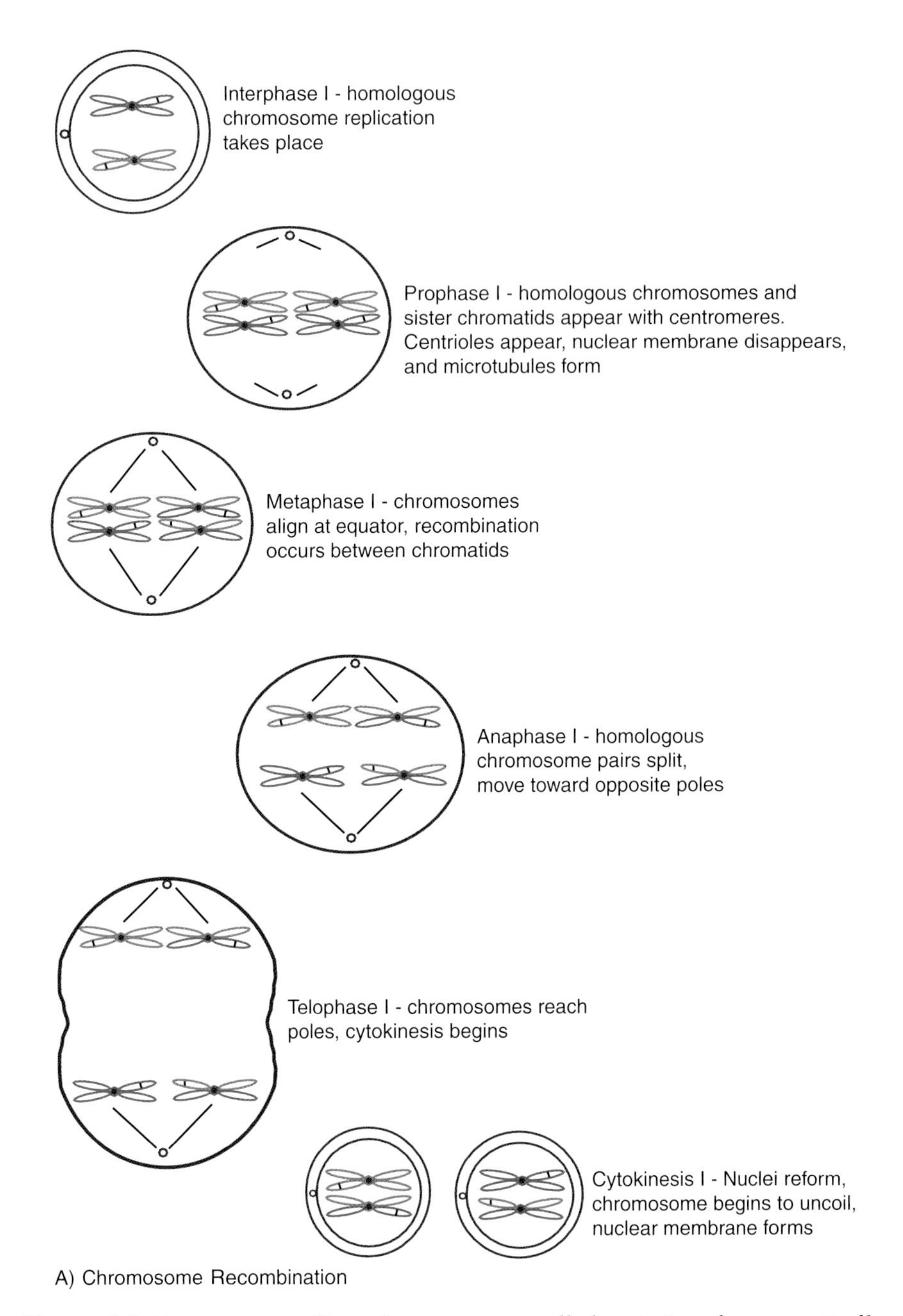

Figure 5.3 Egg or sperm cells undergo a process called meiosis to form genetically unique cells that will later undergo fertilization to form a new organism. During the process of meiosis, chromosome recombination is responsible for generating new DNA combinations and a reduction division occurs so that each sperm or egg cell has a 1N or haploid state. When the new organism is formed, it will be 2N or diploid (1N sperm + 1N egg = 2N zygote).

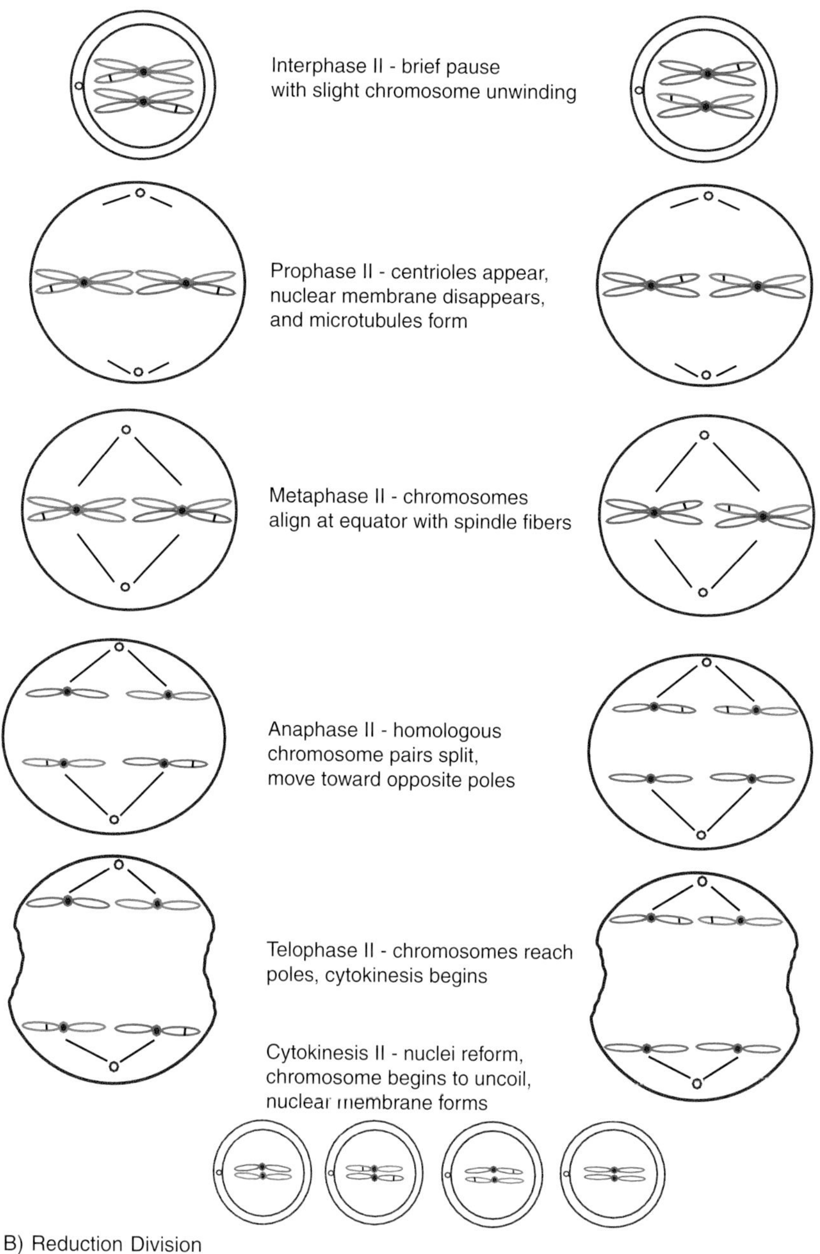

B) Reduction Division

Figure 5.3 (continued)

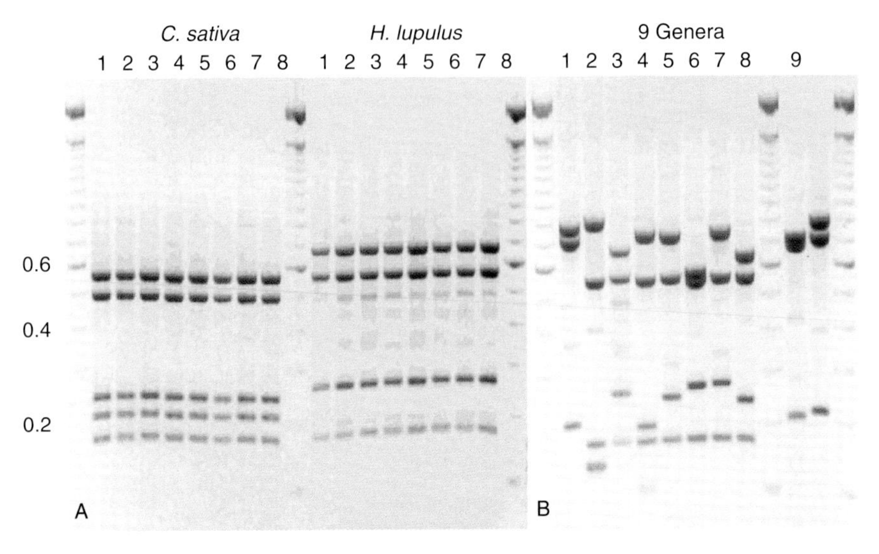

Figure 10.3 Restriction fragment length polymorphism analysis of PCR-amplified chloroplast fragment HK. Panel A (left) compares *Hin*F I RFLP profiles from eight different individuals of *C. sativa* collected from Tennessee and Kentucky and seven different varieties of Hop (*H. lupulus*) (lanes #1–8 are varieties Mt. Hood, Willamette, VT wild type, Liberty, Aquila, Chinook, Fuggles-1, and Fuggles-2, respectively), marijuana's closest relative. Panel B (right) compares the *Hin*F I RFLP profiles from individuals representing nine different plant genera (lanes #1–9 are *Citrus, Cirsium, Rubrus, Lycopersicon, Ficus, Urtica, Morus, Pyrus, Menthe,* and *Syringa*, respectively). Size markers are Gibco BRL 100 bp DNA ladder (Life Technologies Inc.) and size labels are in kilobases. Electrophoresis was carried out in a 1.5% agarose gel and 0.5x TBE buffer. Detection was by ethidium bromide staining and ultraviolet transillumination. Shown are the inverted gray-scale gel images.

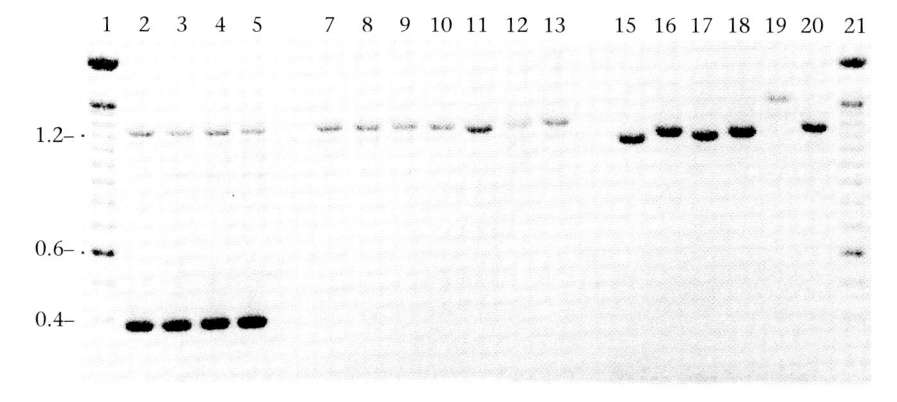

Figure 10.4 PCR co-amplification of the marijuana-specific fragment CS47-50 and the nonspecific chloroplast fragment SfM. Amplification of the 392 bp marijuana-specific fragment occurs only when marijuana DNA is present (lanes #2–5). Lanes #7–10 are *H. lupulus*, #11–13 are *Urticales*, #15 is mint, #16 is tomato, #17 is lilac, #18 is grapefruit, #19 is bull thistle, and #20 is *H5 H. lupulus*, respectively. Amplification was performed with a Stratagene RoboCycler for 35 cycles of 94°C for 1 min., 60°C for 1.5 min, and 72°C for 2 min. PCR reactions were 25 µL and contained 3 m*M* $MgCl_2$, 200 µ*M* dNTPs, 2% formamide, and 1 unit Biolase Taq polymerase in 1x Biolase PCR buffer. Primers CS47 and CS50 were at 200 µ*M*. When SfM primers were included as a positive check on amplification, both primer pairs were present in a mole-equivalent ratio of 20:1 CS primers to SfM primers. Size markers are 100 bp DNA ladder (Invitrogen) and size labels are in kilobases. Electrophoresis was carried out in a 1.1% agarose gel with 0.5x TBE buffer. Detection was by ethidium bromide staining and ultraviolet transillumination. Shown is the inverted gray-scale gel image.

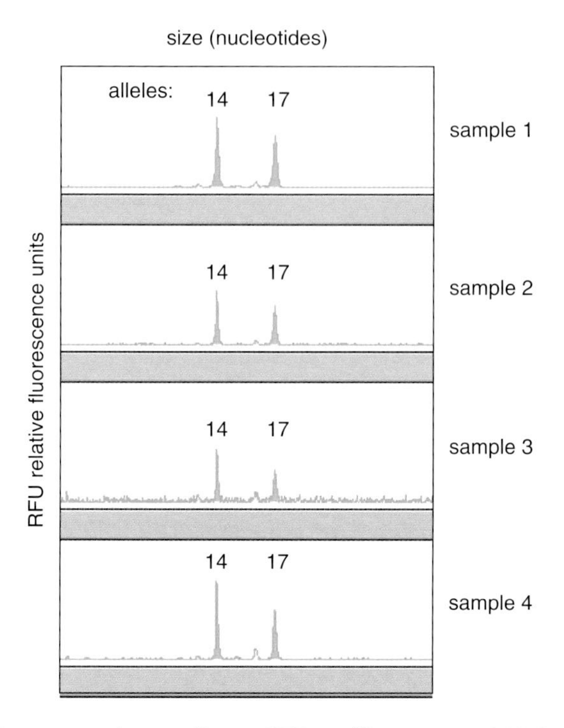

Figure 10.6 Representative marijuana STR profiles run on a 310 Genetic Analyzer show the results of a single locus from casework samples. Four samples are shown, each exhibiting the same STR profile (type 14, 17). The exact genotype was confirmed when compared to an allelic ladder. The x-axis is size (nucleotides) and the y-axis is relative fluorescence units (RFU).

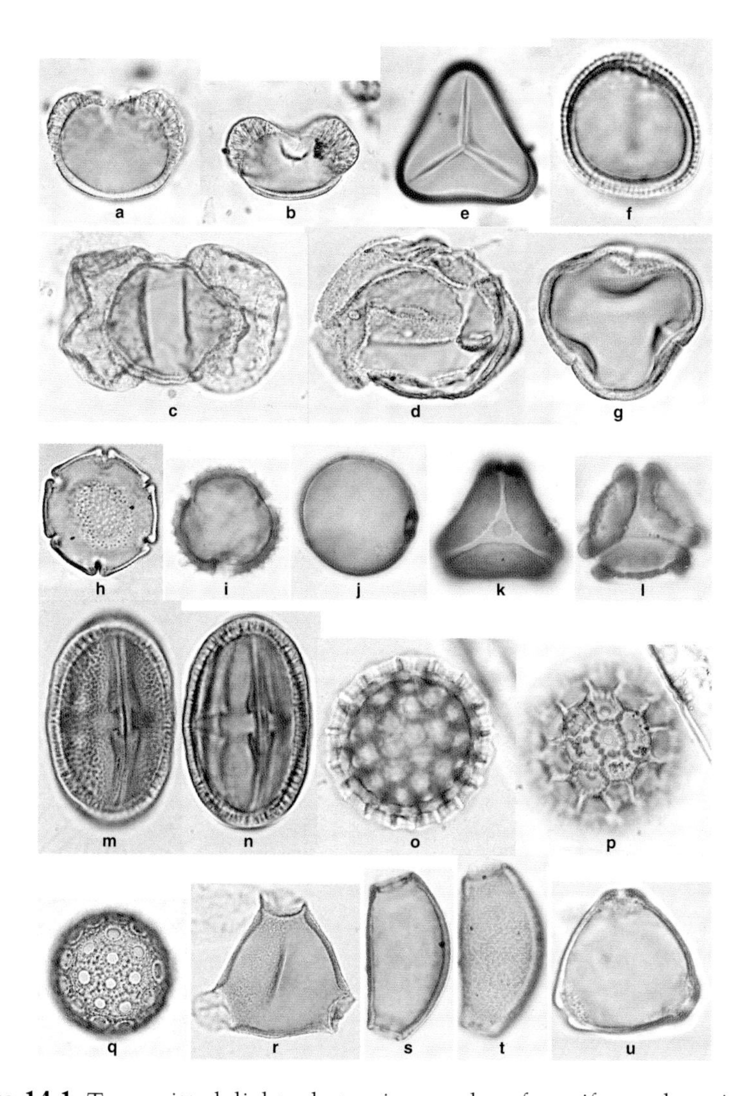

Figure 14.1 Transmitted light photomicrographs of conifer and angiosperm pollen from New Zealand and Australia, and a fern spore. Scale bars = 10 μm. Magnifications range from 600X to 900X. a–d, Conifers; a. *Dacridium*; b. *Halocarpus*; c. *Podocarpus*; d. *Agathis*, e. Fern spore; Cyathea. f–u, Angiosperms. f. *Ascarina* (Chloranthaceae); g. *Coprosma* (Rubiaceae); h. *Nothofagus* (Fagaceae); i. *Olearia* (Asteraceae — "daisy"); j. grass (Gramineae); k, l. Myrtaceae — e.g., *Eucalpytus*; m, n. *Scaevola* (Goodeniaceae), same grain, high and median foci respectively; o, p. *Tribulus* (Zygophylaceae), same grain, median and high foci respectively; q. *Ptilotus* (Amaranthaceae), high focus; s, t. *Banksia* (Proteaceae), same grain, median and high foci respectively; u. *Casuarina* (Casuarinaceae). Photography: a, b, d–h, D. Mildenhall; c, i–u, L. Milne.

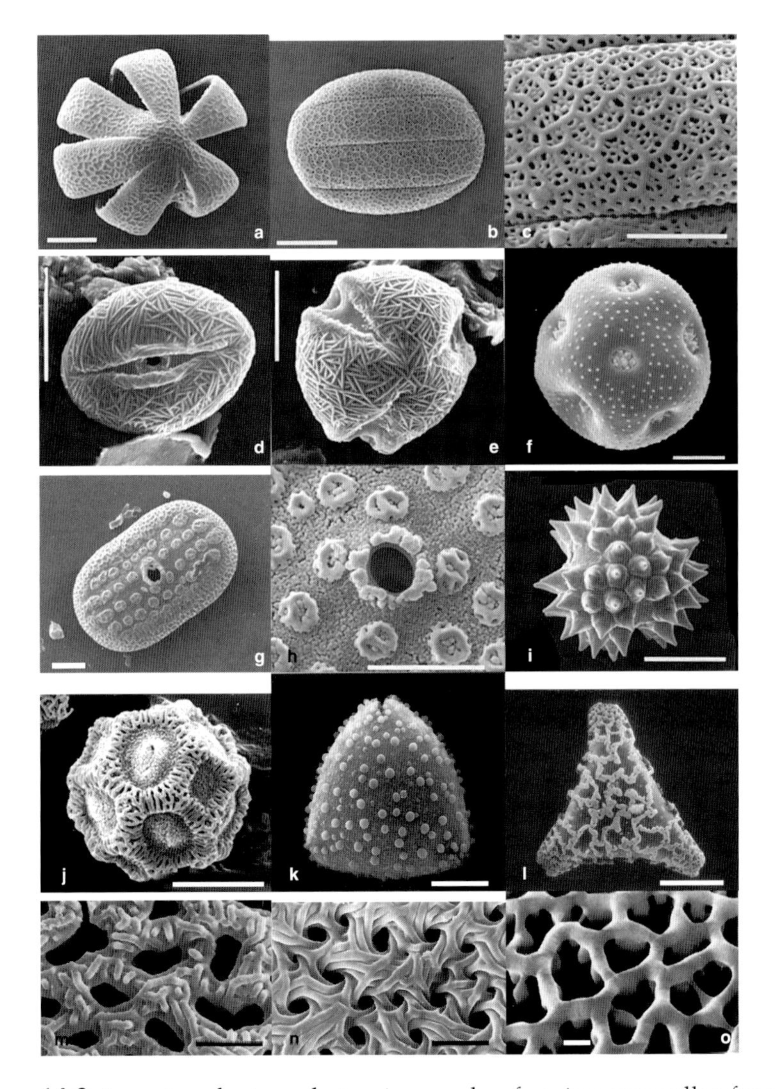

Figure 14.2 Scanning electron photomicrographs of angiosperm pollen from the southeastern U.S. and Australia. Scale bars: a–i = 10 μm, l = 50 μm; m–o = 1 μm. Magnifications range from 700X to 11,500X. a. *Lamium amplexicaule* (Lamiaceae); b, c. *Salvia azurea* (Lamiaceae); b. whole grain; c. surface ornamentation; d, e. *Sedum nuttallianum* (Crassulaceae), equatorial and polar views respectively; f. *Minuartia drummondii* (Caryophyllaceae); g, h. *Justicia runyonii* (Acanthaceae); g. whole grain, h. surface ornamentation; i. *Solidago canadensis* (Asteraceae); j. *Opuntia stricta* (Cactaceae); k. *Beauprea montis-fontium* (Proteaceae); l. *Proteacidites reticularis* (fossil Proteaceae); m–o, Surface ornamentation of three species from the same Proteaceae genus; m. *Petrophile media*, n. *Petrophile conifera*, o. *Petrophile diversifolia*. Photography: a–j, adapted from Jones, G.D., Bryant, V.M., Jr., Hoag Lieux, M., Jones, D.J., and Lingren P.D., *Pollen of the Southeastern United States,* American Association of Stratigraphic Palynologists Foundation, Contributions Series, 30, AASP Foundation, 1995; k–o, L. Milne.

traps and later processing techniques are not designed to remove any attached pollen grains; thus, the drugs retain some plant or geographic history.

14.7.5.2 Case History: Traveling Cocaine

A shipment of 500 grams of cocaine was seized in New York. Pollen analysis was carried out in the hope that it might provide clues about its geographic origin and shipment route. Three distinctly different groups of pollen were present in the sample. One pollen group represented plants growing in Bolivia and Colombia, which were undoubtedly mixed with the sample when the coca leaves were picked and then processed to form coca paste. The second pollen group came from trees that only grow in a few regions of North America — the cocaine must have been cut and packaged at one of these locations. The last pollen group represented weedy plants that commonly grow in vacant lots in New York, where it was probably cut again, packaged, and distributed.[11]

14.7.6 Packing Materials

Materials used in packaging, such as shredded paper, straw, and cardboard, can be useful as pollen traps if they are exposed to the air at their place of origin (outdoors or in open-air warehouses).

14.7.6.1 Case History: Soiled Machinery

A European company exported some machinery to a location in Asia by sea. The boat stopped at a number of ports on the way where various cargo was loaded and unloaded. The machinery was packed in wooden crates, but upon reaching their destination the crates contained only bags of soil. Pollen analysis was conducted on the soil. The pollen and spores recovered were from plants normally found in South Africa. Since the boat had stopped in Cape Town, this narrowed the search area to one port. The missing machinery was discovered several months later in a South African warehouse.[8]

14.8 Comparison of Pollen Assemblages from Control and Exhibit Samples: An Example

Table 14.4 compares the control scene and regional pollen assemblages from an actual crime scene in a *Banksia* woodland, with the pollen assemblages collected from two different parts of a car that was suspected to have been at the crime scene. As noted previously, a pollen assemblage from a vehicle will contain a combination of pollen from plants that occur where a vehicle is commonly used or housed, and from plants where it has most recently been used. Therefore, if the car had been at the crime scene, the pollen assemblages

Table 14.4 Comparison of Pollen Assemblages from Control Soil Samples and Exhibits from a Car

Pollen	Control Samples		Forensic Samples	
	Scene	Regional	Exhibit 1	Exhibit 2
1. *Casuarina*	34.4	36	9.6	8
2. Myrtaceae	33.6	34	38.4	44.8
3. Pine	1.6	0.4	11.2	4
4. Grasses	17.2	7	26.4	31.2
5. *Banksia* 1	2	2.2	0.4	0.4
6. *Banksia* 2	0.4	1.8	0.4	2.8
7. *Stirlingia*	0.4	0.4	0.4	0.8
8. Acacia	—	1	0.4	0.4
9. Other — Gen et sp. indet.	10.4	17.2	12.8	7.6

collected from it will not be identical to those from the control samples, but they may contain some rare species that also occur in the control samples

The considerable differences in percentage occurrence of the wind-pollinated taxa (1 to 4) in the control and scene samples is to be expected, and is not of great significance. However, the pollen assemblages from the evidentiary samples have some pollen species in common with the assemblages from the control sample, the same *Banksia*, *Stirlingia*, and *Acacia* species, and several "other" species (not listed). This suggests that the car had either been in a region with very similar vegetation or had been at the actual crime scene. This case has now been resolved. The car had indeed been at the crime scene, and was then used to transport a homicide victim to a pine plantation. An excess of pine pollen in tape lifts from under the vehicle and the higher percentage of pine pollen in other evidentiary samples indicated where the body might be located.

14.9 Sample Collection and Storage

Even if palynology is not considered necessary during the initial investigation, it is wise to collect control samples from the crime scene and store them in case pollen analysis may be required later. Ideally, forensic palynology samples should be collected and handled only by a forensic palynologist. Because this is not always possible, and because other procedures may need to be conducted on the sample prior to pollen analysis, it is important that the palynologist is informed of the method of collection and what procedures have been carried out on the sample prior to pollen analysis.

The most critical aspect of collecting samples for forensic palynology is to ensure that there is no contamination during collection and storage. Because pollen is plentiful in the atmosphere and settles on any exposed surface, improper collection and storage and accidental contamination of forensic samples not only leads to inaccurate results, but may be grounds for the dismissal of any pollen evidence in court. Fortunately, contamination-free samples can be validated if the history of the sample is well documented.

Many of the procedures used to guard against pollen contamination of the evidentiary sample may already be routine procedures used for the collection of other types of forensic samples, but there are certain precautionary measures that are particularly important to ensure accurate and valid pollen analysis. First, all sampling and any sample to be examined for potential pollen grains should be handled with new and sterile disposable gloves. All implements used to collect the forensic pollen samples should be thoroughly cleaned before use and stored in sealed, airtight containers until needed for use. Once samples are collected, they should be stored in sterile, airtight containers and sealed immediately. Once sealed, the samples should not be opened until they are ready to be examined by a palynologist. Other precautionary practices are discussed more fully below.

14.9.1 Pollen Control Samples

Control samples should be collected methodically and with the purpose of providing a baseline pollen record for later comparison with pollen assemblages recovered from items of forensic interest. Knowing the best number of control samples to collect from the scene of a crime or from another locality that is deemed forensically important may be difficult. In general, the greater the number of control samples that are collected and examined, the more pollen information will be available for a given locale. Because the pollen in each control sample may vary slightly in the plant types and percentages of pollen grains found, their combined record offers a potential range of pollen variation for a given location. This knowledge allows for matching the pollen results from forensic samples with actual locations with a greater degree of certainty. This is especially true when it can be shown that the pollen types in a forensic sample fall within the range of expected pollen variations of the control samples. Not having sufficient control samples to examine, or having only one control sample, weakens the certainty of a forensic pollen sample if it contains a pollen assemblage similar to but not exactly like the pollen data from a single control sample. It is better to have too many control samples than not enough, and decide at a later date whether to examine all or only some of the samples.

14.9.2 Collection of Pollen from Soil, Mud, and Water Samples

Very small soil or mud samples are often adequate for pollen analyses, but what is meant by "small?" Ideally, 15 to 20 g of soil or mud should be collected to ensure that there are enough samples to perform all tests and to retain some of it for future reference. Soil is best collected and stored in sterile, airtight glass or plastic containers or sterile zip-lock-type plastic bags. Specimen jars or plastic centrifuge tubes with screw tops are also useful collection and storage containers. Plastic screw-top containers with a built-in scoop in the lid are ideal. Often known as "pooper scoopers," pathologists use these types of containers for feces collection. Plastic bags are not ideal for all types of forensic samples because of static, and it is preferred that cloth bags are not used because pollen can travel in and out of the woven surfaces. For water, one to two pints (500 mL to 1 L) is often sufficient. Like other samples, water should be stored in a sterile and sealed container.

When persons committing a crime travel through an area, their vehicles, personal belongings, clothing, or tools may collect ambient pollen from the air or pollen from surfaces they come in contact with. This is one reason why pollen control samples taken from the immediate crime scene and its surrounding area are of critical importance. Past research has shown that the most effective method of collecting each control soil sample requires the selection of 10 to 20 regularly spaced "pinch samples" (one half to a teaspoon full) from the top one centimeter of the soil by walking back and forth in the selected sampling area. All of the "pinch" samples from one control sample location are placed into the same sterile container. The container is then sealed and shaken lightly to ensure that the samples are thoroughly mixed.

Generally, one or more scene, locality, and regional control soil samples are collected, respectively. Scene or site samples are taken from an area approximately 1 to 2 m^2 that includes the crime scene. Locality samples are usually collected from an area 50 to 100 m^2, surrounding and including the crime scene. If the plant associations in an area are not consistent, several types of regional samples may be taken. An overall regional sample can be taken by combining samples taken from the four cardinal directions 50 to 100 m away from the crime scene, or in directions where the plant communities are different to those observed at the crime scene. Alternatively, four separate samples may be collected from each of these directions, or separate samples can be taken starting at the crime scene, and along a transect in each cardinal direction to a distance of 50 to 100 m away from the crime scene. Samples may also be collected along other transects where it is suspected a person or vehicle may have traveled to or from the crime scene.

Water from puddles, ponds, streams, lakes, rivers, and the ocean, as well as any open containers, will contain ambient pollen and pollen from water runoff. Control water samples should be collected from the edge of any bodies of water, and when possible they should also be collected a few meters away from the shoreline. At the same time, mud along the edge of the water body, and when needed, the bottom of the water body, should also be collected.

14.9.3 Storage of Soil, Mud, and Water Samples

Storage procedures should be the same for both the control samples and potential forensically relevant pollen samples (evidence). Correct storage procedures are essential to prevent the samples from being altered or contaminated by microbial activity. Certain bacteria and fungi feed on pollen, and their mycelia can make pollen analysis significantly more difficult.

Pollen samples that are completely dry can be stored indefinitely at room temperature. Moist or damp samples should be refrigerated or frozen to prevent any microbial growth. For water and sludge samples, a small amount of alcohol or phenol should be added before the sample is refrigerated. If a sample is to be stored for a long period of time prior to pollen analysis, damp soil samples may be dried individually in a special drying oven. All soil and water samples may be frozen, as freezing will not damage either pollen or spores.

14.9.4 Plant Reference Samples

To help identify the pollen types found in a control sample, the vegetation at a site of interest should be thoroughly surveyed and photographed by a palynologist or botanist. If this is not possible, samples of dominant plant types should be collected for formal identification by a reputable botanist. Plants that are flowering at the crime scene are especially important to note because the person or persons committing a crime may have brushed against them and carried pollen on their clothing. Pollen from these plants may also dominate a pollen assemblage from soil at the immediate site. However, it is important to remember that pollen from plants at the crime scene that are not flowering at that time may also be present in the soil at varying depths and differing abundance.

14.9.5 Collection and Storage of Plant Reference Specimens

If a palynologist or botanist is unable to attend a crime scene at the initial investigation, it is important that crime scene personnel are aware of how much of a plant to collect and how to properly store plant samples to preserve them until they can be formally identified by a botanist. Small plants (less than 30 cm in height) may be collected intact, but for larger plants a representative section (at least 30 cm long) that includes leaves on a stem, and flowers or fruit if available, should be collected. The collected plant samples should be placed in cardboard folders, large envelopes, or between sheets of newspaper or other absorbent paper. They should never be placed in plastic or sealed nonabsorbent containers as they will wilt or become moldy and potentially difficult to identify. Plant samples may be refrigerated if they are to be stored for short periods of time, but if they are to be stored for more than a few days they should be dried and pressed. Pressing a plant specimen removes moisture and preserves its identifiable features. For example, dried plant specimens are kept in herbaria for hundreds of years. To press plant samples, they are placed between several layers of newspaper or other absorbent paper or cardboard, placed one on top of the other, and weighted down with a heavy object. Specialized presses that strap samples between two sheets of wood are often taken to a site by the botanist or palynologist. Most plants dry in less than a week, but while they are drying it is important that the press is well ventilated and the paper changed several times to prevent mold growth.

14.9.6 Additional Collection Tips for Palynology

It is important that all collection implements are thoroughly cleaned and stored in a sterile container or plastic bag prior to use. Obvious soil, mud clods, and plant material can be removed from evidence by scraping with a sterile knife or brushes. Soft paintbrushes, makeup brushes, and toothbrushes are ideal tools for this purpose. The collected material may be scraped or brushed onto aluminum foil that is then folded and sealed, or brushed directly into a sterile storage container. Fine dirt particles and dust may be washed off, vacuumed, or removed with tape lifts.

Crime scene investigators routinely use tape lifts to collect fibers and soil for analysis. Tape lifts are especially useful for palynology if the material needs to be preserved in a sequence, or if there is not enough material for all forensic disciplines to have a separate sample for study. Tape lifts are usually taken with print-size sheets with one adhesive surface that, after lifting is completed, can be covered by a grease-proof protective sheet to preserve the evidence and guard against possible contamination. Other types of adhesive tapes may be used in the laboratory or in the field. If a protective covering is not available, a tape may be folded over onto itself or placed adhesive side down onto a clean plastic surface (e.g., plastic or Styrofoam® plates, lids, or containers). The use of tape lifts is not recommended if only pollen analysis is to be done, or no other method of collecting is possible, as this prolongs the costly processing time. Vacuuming is useful for removing pollen grains from large surfaces such as floors, carpets, and the interior of vehicles. Small handheld vacuums that have been thoroughly cleaned can be used with a clean filter or bag for each sample.

Because pollen is microscopic, it is often present on surfaces and in clothing and other woven goods that may appear to be clean. Tape lifts and vacuuming, as previously discussed, will remove some microscopic debris, but washing is the most effective method to remove all dust and other microscopic particles. Evidentiary items may be soaked in distilled water and detergent, or in a strong surfactant such as Calgon®. This should take place in a clean container with a lid that can be sealed. For large items with dense fibers such as carpets, soaking with regular rotation in the soaking medium over several days will generally remove even firmly ingrained particles including pollen. After soaking, the evidentiary item is removed and the remaining water is either centrifuged to concentrate the residue, or allowed to stand for several hours until the residue settles to the bottom of the container and the water can be carefully removed from the pelleted debris. When the latter technique is used, alcohol should be added to the water solution to reduce its specific gravity below 1.0. Research has shown that some types of insect-pollinated plants have pollen containing lipids and other types of oils, and large bisaccate pollen grains containing minute air bubbles will remain afloat in water regardless of the amount of time the materials are allowed to settle.

14.9.7 The Use of Multiple Collection and Extraction Methods

More than one type of palynological sample may be taken from evidence to provide information about its whereabouts during and prior to a crime being committed. Shoes are an excellent example for illustrating how multiple collection methods can provide useful information to the scientist and crime scene investigator. If the shoes are collected shortly after a crime has occurred, any obvious surface material that is scraped from them will likely contain a pollen assemblage from their most recent use — for example, the crime scene or an alibi location. Washing the surfaces of shoes will yield a combination of pollen types from their latest use (crime scene or alibi location) as well as from places they may have been worn since they were last cleaned. This is particularly important to note when shoes thought to belong to a person who committed a crime are found at the scene. If crime scene investigators have no suspect in mind, then the pollen assemblage from the shoes or other items at a crime scene might provide a clue as to where the person lives and works (see the case history, "Cereal Points to a Serial Rapist").

Automobiles are another example illustrating the use of different sample extraction methods to provide useful historical information. Ambient pollen and pollen in dust will collect in air cleaner filters and radiators. Pollen collected from an automobile's most recent destinations can usually be removed by gently shaking or brushing of the air filter or radiator, but pollen that has been in the vehicle for some time will probably be incorporated in an oily or greasy residue that can only be removed by solvents. Washing in hexane will remove pollen from that sort of residue such as might be found in the engine compartment.

14.10 Examination of Forensic Samples

Ideally, a palynologist should remove forensic pollen samples from exhibits in a palynology laboratory. However, rarely do forensic personnel have access to a specialized facility for pollen analysis. Typically, pollen analysis would be performed by a qualified academic pollen identification expert or by a trained trace analyst at a crime laboratory. For smaller items that can be transported in sealed exhibit bags, such as clothing, ropes, baskets, rugs, shoes, and smaller parts of vehicles that can be detached, it is easy to transport and analyze evidentiary items at the expert's laboratory. For larger items, it may be necessary for forensic pollen samples to be collected at the crime scene by a palynologist or crime scene personnel. Wherever the samples are collected, it is crucial that great care be taken to guard against contamination from airborne pollen. Some evidence may need to be examined by other qualified forensic specialists (e.g., soil scientists, entomologists, chemists, pathologists, fiber experts) before a palynology sample can be collected. In such circumstances, it is important to have a basic understanding of forensic pollen analyses in order to protect the quality of the sample, much like an evidence examiner would take great care in removal of a bloodstain without destroying possible latent prints on the evidentiary item.

Forensic palynologists work in contamination-free laboratories and take special precautions to ensure against contamination from glassware or during sample handling. The palynology samples and the evidence they were taken from are kept in a security safe or other locked equipment when they are not being handled by the palynologist. This maintains the chain of custody for evidence and enables a palynologist to state under oath the complete history of the samples from the time they received them. Several "test slides" (e.g., slides coated with a sticky substance such as glycerol, glycerine jelly, Vaseline®, or double-sided adhesive tape) can be placed around the laboratory during sample collection and pollen preparation as another precautionary measure in the event that errant ambient pollen that might contaminate a sample is present in the room. Test slides are especially important if a forensic sample is collected in premises other than a palynology laboratory.

14.11 The Future of Forensic Palynology

The potential benefits of using forensic palynology are unknown to most police and law enforcement agencies, lawyers, and other forensic scientists. Bryant et al.[3] discussed in detail some of the reasons why they believed this field of forensics to be one of the "least known and least attempted techniques." They noted that, at that time, fewer than 10% of police laboratories realized pollen might be useful as evidence, and even fewer had any knowledge of cases in which palynology had been used. More than 10 years later, forensic palynology is now becoming more widely recognized, but its use in forensic casework is still uncommon. As with many specialty areas of forensic testing, new methods for evidentiary analyses are not widely used or accepted when first developed. Worldwide acceptance of new technologies as major forensic tools has often resulted from the promotion of the technique by the scientists themselves, and by the ever-increasing number of courts that accept their findings as valid and useful evidence for the resolution of criminal and civil cases. Palynology as a tool to help resolve crime has a bright future; however, its level of use will depend on both the proper education of crime scene personnel and forensic scientists, and the training of future palynologists dedicated to working in this field.

References

1. Erdtman, G., *Pollen Morphology and Plant Taxonomy: Angiosperms — An Introduction to Palynology*, E.J. Brill, Leiden, 1986.
2. Wodehouse, R.P., *Pollen Grains*, McGraw-Hill, New York, 1935, 6.
3. Bryant, V.M., Jr., Mildenhall, D.C., and Jones, J.G., Forensic palynology in the United States of America, *Palynology*, 14, 193, 1990.
4. Faegri, K., Kaland, P.E., and Krzywinski, K., *Textbook of Pollen Analysis*, 4th ed., John Wiley & Sons, New York, 1989, chap. 11.
5. Tauber, H., Differential pollen dispersion and filtration, in *Quaternary Paleoecology*, Cushing, E. and Wright, H, Eds., Yale University Press, New Haven, CT, 1967,131.
6. Wrigley, J.W. and Fagg, M., *Banksias, Waratahs & Grevilleas, and all Plants in the Australian Proteaceae Family*, Harper Collins Publishers, London, 1989, 82.
7. Horrocks, M. and Walsh, K.A.J., Forensic palynology: assessing the value of the evidence, *Rev. Palaeobot. Palynol.*, Special Issue: New Frontiers and Applications in Palynology, 103, 69, 1998.
8. Jarzen, D.M. and Nichols, D.J., Pollen, in *Palynology, Principles and Applications*, Jansonius, J. and McGregor, D.C., Eds., American Association of Stratigraphic Palynologists Foundation, Publishers Press, Salt Lake City, UT, 1996, chap. 9.
9. Mildenhall, D.C., Forensic palynology in New Zealand, *Rev. Palaeobot. Palynol.*, 64–65, 227, 1990.
10. Mildenhall, D.C., A grain of evidence: invisible pollen grains can be enough to sort out the guilty and the innocent, *New Zealand Science Monthly*, June 1998.
11. Stanley, E.A., Applications of palynology to establish the provenance and travel history of illicit drugs, *Microscope*, 40, 149, 1992.

Additional References

Bryant, V.M., Jr. and Mildenhall, D.C., Forensic palynology: a new way to catch crooks, in *New Developments in Palynomorphs Sampling, Extraction, and Analysis*, Bryant, V.M. and Wrenn, J.W., Eds., American Association of Stratigraphic Palynologists Foundation, Contributions Series, 33, 145, 1998.

Erdtman, G., The acetolysis method. A revised description. *Svensk Botanisk Tidskrift*, 54, 561, 1960.

Milne, L.A., Forensic palynology — cereal points to serial rapist, in *Program. The Crim. Trac. 15th Int. Symp. on the Forensic Sciences*, the Australian and New Zealand Forensic Society, 2000, 106.

Playford, G. and Dettmann, M.E., Spores, in *Palynology, Principles and Applications*, Jansonius, J. and McGregor, D.C., Eds., American Association of Stratigraphic Palynologists Foundation, Publishers Press, Salt Lake City, UT, 1996, chap. 8.

Stanley, E.A., Forensic Palynology, in *Proc. Int. Symp. on the Forensic Aspects of Trace Evidence*, Washington Government Printing Office, 1991,17.

Steussy, T.F., *Plant Taxonomy*, Columbia University Press, New York, 1990, chap.18.

Traverse, A., *Paleopalynology*, Unwin Hyman, Boston, 1988.

14.12 Case History: Cereal Points to a Serial Rapist

LYNNE A. MILNE

Police suspected that the same person committed five sexual assaults that had occurred over a period of 5 years in the metropolitan area of Perth, Western Australia. The only evidence they had was the rapist's DNA, an Identikit picture compiled from various victims' accounts, and a pair of track shoes that they thought belonged to the rapist. The shoes had been left in the car the rapist had stolen from a northern suburb, and the car was used to travel approximately 30 km to where he committed a sexual assault. On leaving the scene, he took a corner too sharply and had an accident. The rapist abandoned the car and escaped into the surrounding bush land. The track shoes found in the car did not belong to the person who owned it, or any of their family, and did not belong to the victim. This particular assault occurred early in the second year of the rapist's activities, and at that time, the Western Australian Police Service had no knowledge of forensic palynology and how it might help their investigation. Three years later, a small and select group of police officers attended a presentation on forensic palynology. The palynologist was asked if pollen could remain on shoes that had been stored in a sealed exhibit bag for three years. If so, would it be possible to determine where the rapist lived or worked?

Because of the intermittent timing and the broad geographic spectrum of the sexual assaults, police suspected that the rapist might be an itinerant worker that worked in the country and came to the city occasionally. From other evidence, they also knew he was of Australian aboriginal descent. Periodic visits to prisons throughout the state to collect DNA from possible suspects had been unsuccessful in resolving the case. Small amounts of soil and dust were recovered from the soles of the track shoes, and numerous fragments of plant material that were lodged firmly in the outer edge of the soles were removed. Pollen analysis of this evidentiary material revealed a rich pollen assemblage consistent with Eucalpyt woodland vegetation. The assemblage was dominated by pollen from gum trees, gum tree relatives, and grasses. A small percentage of pollen from sheoaks and pine trees was also present, and a number of unidentified pollen types also occurred. Eucalpyt woodland is the most common natural vegetation of the southwest of Western Australia. It can also be found in areas 100 km or more from the city to the north and east as well as in city parks, golf courses, and in the area that the assault occurred. The pollen assemblage narrowed the search area down to one third of the state but also to the area of densest population. Police urged for the identification of the unknown pollen types to see if they might further refine the search area.

The brittle fragments of plant material deeply lodged in slits in the soles of the shoes were investigated further. One of the larger unidentified pollen types that comprised 1% of the pollen assemblage had the appearance of large grass pollen (round with one pore) and proved to be from a cereal plant (wheat, oats). Cereals are grasses and they are cleistogamous. Pollination occurs inside the unopened flower so they do not produce very much pollen, and their pollen is rarely found in soil. The strawlike fragments were difficult to dislodge from the shoes because they were exactly the same lengths as the slits. Discussions with farmers confirmed that stubble (the short, upright, very brittle straw left in a field after hay cutting) could break away from the plant and lodge in the tracks of shoes as a person walked over it. Wheat pollen was collected from the experimental crops of the local Agriculture Department and from local farmers. The wheat pollen was a good visual match with the large, grasslike pollen grains found in the pollen assemblage from the shoes. Together, the straw and wheat pollen suggested the rapist had walked over crop stubble and may have been involved in cutting hay. Although many farmers cut hay for their own local use, hay cutting is most common in the chaff-cutting industry. This industry grows wheat especially to be cut and sold as chaff for livestock fodder. The chaff-cutting industry in Western Australia employs teams of itinerant workers, who are predominantly of aboriginal descent, to process the hay for drying in November and December and cut the hay into chaff during January and February. The track shoes were collected after an assault in early March.

Chaff-cutting farms occur in a small area some 60 to 100 km from the Perth metropolitan area. Police approached the farmer with the largest farm to collect the names of workers he employed. Less than a week later, police arrested a man who confessed to the sexual assaults. He was related to a person employed by the farmer, and he worked in a business that provided services to the chaff-cutting industry. This case is an example of how forensic palynology can assist in profiling the personal belongings of a suspect to aid a police investigation. A case that had puzzled police for years was solved in a few short weeks.

14.13 Case History: Hobbled by *Hypericum*

DALLAS C. MILDENHALL

A woman was asleep in her bedroom in a northern New Zealand coastal town, but had left the back door unlocked because she was waiting for her live-in boyfriend to come home. She was disturbed in her sleep and, thinking that her boyfriend had come home, awoke to find a stranger in her bed fondling her breasts and someone watching from the foot of her bed. When she screamed, the assailants ran off but not before she had grabbed the more forward assailant and wrestled his denim jacket off him. The jacket was left on the kitchen floor.

In the meantime, the boyfriend arrived home to find his partner on the telephone to the police. At the same time one of the perpetrators returned and knocked on the door. Seeing his coat on the floor he asked for it back, saying that he had left it during an earlier visit. The boyfriend, not knowing what had happened, did not stop the man from retrieving his jacket. As he hurriedly left, the perpetrator brushed the jacket against a flowering *Hypericum* bush growing outside the kitchen door. It was then discovered that a wallet and purse from the kitchen table were missing. The man who retrieved the jacket was soon found and arrested by the police. He denied any knowledge of the incident even though the victim and her boyfriend identified him.

Hypericum pollen occurs infrequently in forensic samples in New Zealand, but never in more than trace quantities (1% or less). This was a case in which more than trace amounts of *Hypericum* pollen must occur on the clothing of the suspect, especially the jacket, which had been in direct contact with the flowers at the height of their pollen production. The denim jacket contained 24% *Hypericum* pollen, but not only that, his pants contained 14% and his polo shirt 27%, respectively. To have these percentages, it was likely that all the items had been in direct contact with the bush. Traces of *Hypericum* pollen were also found on his shoes and socks. These trace amounts could have been picked up when the man brushed past the bush, or by transference from his outer clothing. The pollen still had the cell contents preserved and a number of grains were in large clumps that could not have been aerially dispersed, suggesting recent and direct contact with the bush. The suspect was charged with sexual assault on a female and burglary, and found guilty and sentenced appropriately.

14.14 Case History: *Nothofagus* Noticed

A man was found shot in the back on Mount Holdsworth in the Tararua Ranges near Wellington, the capital city of New Zealand, which is situated on the southern coast of North Island. The vegetation around the body consisted in part of a number of silver beech trees (*Nothofagus menziesii*), a species that is very rare in Wellington. Pollen from this species does not disperse well and, even under windy conditions, falls within a few kilometers of its source. Pollen analysis of the victim's clothing indicated that *N. menziesii* pollen was a common type. Since the murder took place well into the forest, off the well-known tracks, the chances were that the person who committed the murder would also have traces of this pollen type on his or her clothing. Police investigations went on for 18 months before a man from Wellington was arrested for the murder. He denied having been in the mountain region.

Approximately 60 items belonging to the suspect and his associates were examined for pollen. Of these items, four were found to contain significant numbers of silver beech pollen, and eight others contained 1% or less of silver beech pollen. Significantly, 4% of silver beech pollen was identified on a blue Fairydown jacket, and 2% on a pair of tricolor Rip Curl board shorts. When confronted with this evidence, the suspect maintained that he had purchased the Rip Curl shorts after the time of the murder, in Kaikoura, a town on the northeastern coast of the South Island of New Zealand. He also asserted that he had purchased the Fairydown jacket overseas, and that he had never worn it outside the Wellington metropolitan area. Investigations showed that the suspect could not have purchased the Rip Curl clothing from Kaikoura because wholesalers had never shipped those clothing brands to that area. Furthermore, silver beech did not grow near Kaikoura, or in the area between Kaikoura and Wellington. Pollen analysis of the jacket identified a number of pollen types including silver beech, which suggested the jacket had been out of the Wellington city limits into a mountainous area where silver beech can be found. Pollen analysis of control samples from where the suspect lived, from where his father lived, historical pollen samples from Wellington city, and samples obtained from the foot of the mountains close to where a witness reported having seen the suspect did not contain any silver beech pollen, although pollen from other species of beech were common. Rare silver beech trees are known to be growing in reserves and botanical gardens near Wellington, but the suspect stated he had never been near these areas, and these plants were remote from where he normally lived and worked. It would be anticipated that the pollen from these individual trees would be quickly swamped by the greater numbers of wind-dispersed pollen from emergent trees and grasses growing in the same areas.

A witness had been able to identify a man seen with the victim near the foot of the mountains prior to the murder. He identified him by the distinctive colored Rip Curl clothing and blue Fairydown jacket that he was wearing at the time. Prior to analysis, the palynologist was not told of the eyewitness' statement. Out of the 60 items the palynologist examined for pollen, these two articles of clothing were among the four items found to contain silver beech pollen. During the trial, the evidence that the jacket and Rip Curl clothing had been in the mountains near Mount Holdsworth was accepted, and the suspect admitted that he had been there. However, the jury was unable to accept the circumstantial evidence that he had been involved in the killing, and he was found not guilty of murder and discharged back to prison to serve his time on a previous drug charge.

An important aspect of this investigation was that the murder, and subsequent discovery of the body about three months later, took place after the end of the flowering season for silver beech. Despite this time disparity, the area the body was found in must have been so swamped with silver beech pollen that, well after the flowering season was over, the pollen was still evident and able to be dispersed. Silver beech pollen formed between 6% (right sock and shorts) and 21% (right shoe) of the total pollen assemblage found on the clothing of the victim. Also, there was enough silver beech pollen available to enable the suspect, who was only on Mount Holdsworth for a few hours, to gather a significant number of grains on his clothing.

Classical and Future DNA Typing Technologies for Plants

15

ERIC J. CARITA

Contents

0-8493-1529-8/05/$0.00+$1.50

15.1 Introduction

Over the past 20 years the field of forensic plant genetics has evolved at a rapid pace. From the early use of the Random Amplified Polymorphic DNA (RADP) technique to distinguish variability between and within species to the recent developments of locus-specific short tandem repeats (STRs) and single nucleotide polymorphisms (SNPs), the use of genetic testing within the field of forensic botany has received major attention. What was once thought of as an obscure science has provided important investigative aid within murder, rape, and illegal drug propagation cases. Although some forms of plant genetic testing may never be as discriminating as their human counterpart, the test methods are similar with respect to linking suspects, victims, and crime scenes to each other.

Receiving national attention within such cases as the Lindbergh kidnapping and the Arizona paloverde murder case, recent forensic botany advancements have opened doors that were never thought possible. Today, 20 milligrams (and possibly less) of plant material can be efficiently extracted using mechanical cell-disruption techniques to generate DNA profiles that were impossible just a few years ago. Plant evidence that was once overlooked as having little forensic value or relevance is now being collected due to the technological advances within the forensic DNA field. Today, only a few labs perform forensic-based plant genetic typing, but in the future the number will undoubtedly increase due to the need for such a specialized form of testing. Although not new, forensic botany (encompassing both microscopic and plant DNA analyses) is not as commonly utilized as human autosomal, Y-chromosome, or mitochondrial genetic typing. However, the field of forensic botany has shown a presence within the field of applied genomics and is becoming more popular as an increasing number of criminal investigations require its use. This chapter examines classical, present, and future plant DNA typing methods; the advantages and disadvantages of each technique; and their applications in terms of forensic case studies (see Table 15.1).

15.2 Classical Plant DNA Typing Methods

15.2.1 Random Amplified Polymorphic DNA (RAPD) Analysis

Random amplification of polymorphic DNA (RAPD, pronounced *rapid*) is a PCR-based technique used to obtain and compare genetic fingerprints or

Table 15.1 A Comparison of Plant DNA Typing Methods

Genetic Testing Method	Advantages	Disadvantages
Random amplified polymorphic DNA (RAPD)	• No prior genome sequence required • Minimal cost • Amenable to automation	• High molecular DNA is required • Large quantity and pure sample are needed • Can't differentiate single source from mixture samples • Low degree of reproducibility between analysts and laboratories
Amplified fragment length polymorphisms (AFLP)	• Potential to individualize highly inbred plant species • Large number of primer sets allow for a discriminating composite profile • No prior genome sequence required • Amenable to automation	• High molecular DNA is required • Large quantity and pure sample are needed • Can't differentiate single source from mixture samples • Difficult to identify contamination • Counting method results in low discrimination power
Fragment length analysis (STR, VNRT, microsatellites)	• Single or multiplex capabilities • Can still produce a usable profile from low quantity and/or degraded samples • Can determine single source from mixture samples • Fast, automated • Can produce a very discriminating statistic	• Prior genomic sequence required • Difficult to optimize multiplex • Interlab validation and population database is required for each species • Expensive
Single nucleotide polymorphisms (SNP)/microchip technologies (CE Microchip)	• Automated • Thousands of SNPs and multiple STRs per microchip • No PCR artifacts (Stutter, Minus A) • Good for low-level and degraded samples	• Prior genomic sequence required • Interlab validation and population database are required for each species • Expensive • Difficult to differentiate single source from mixtures

profiles within a strain or species. Commonly used for gene mapping studies, the RAPD technique involves the use of synthetic oligonucleotide primers of arbitrary sequence that randomly bind at complementary sequences along the plant genome. Used to amplify both strands of DNA, these 10 to 20 base pair (bp) primers simultaneously amplify anywhere from 3 to 10 polymorphic sites along the DNA sequence.[1] The quality and quantity of amplified product are dependent on the following:

- The length of the primers (primer length will eventually determine the complexity of the profile and the difficulty of analysis)

- The distance between each forward and reverse primer that allows for efficient amplification (optimally <1,500 bp)
- The presence of sufficient variation within the primer-binding regions to effectively differentiate between closely related samples

The amplified product that is resolved by gel electrophoresis and the fragments are compared to determine the extent of genetic variation between samples.[2]

15.2.2 Advantages and Disadvantages to RAPD Analysis

Although the RAPD method is typically not used today, there are a number of advantages that make this technique suitable for forensic plant typing. These include

- No prior genomic DNA sequence knowledge is required for the construction of the oligonucleotide primers.
- The assay is easy and quick.
- The cost of performing RAPD analysis is minimal.
- The RAPD technique allows for the analysis of fragments throughout the organism's entire genome.
- The data produces multiple band patterns, which allows for greater discrimination power.
- RAPD analysis is amenable to automation.

However, with these advantages come disadvantages that can be problematic, especially since analysts do not know *a priori* the condition, age, or degree of degradation that has occurred within an evidentiary plant sample before it reaches the laboratory. One of the major problems that has plagued this technique is that a significant amount of high molecular weight DNA (>30,000 Kb) is required in order to obtain a meaningful profile. This means that the purity and quantity (5 to 50 ng required) of the genomic DNA must be excellent. The issue of contamination between two samples or two species is of greater concern, as compared to more recently developed STR PCR techniques. This is because the short oligonucleotide primers are not locus-specific and may amplify DNA from a large number of genomes, including humans. If contamination has occurred, an inexperienced analyst may not be able to recognize that there is a problem. Therefore, a significant number of quality control measures need to be in place when performing all forms of plant DNA testing, but especially for those that are not species-specific. Finally, and most important when used in court, is the reported low degree of reproducibility between analysts and laboratories.[3] Within the field of forensic botany, the use of RAPD analysis has been limited. However, its full potential was quickly discovered when this profiling technique was called upon to help solve a murder.

15.2.3 Can RAPD Analysis Help Catch a Killer?

In early May of 1992, the body of a deceased black woman identified as Denise Johnson was discovered in Phoenix, AZ. Her nude, dead body was discovered under a paloverde tree (*Cercidium* spp.) and the tire tracks leading from the scene were obvious to investigators. In addition, a pager was found near the woman's body, a number of large gouge marks were evident on the paloverde tree, and scrape marks in the dirt demonstrated signs of an apparent struggle.[4] An eyewitness soon came forward to notify authorities that a white pickup truck had been seen speeding away from the same area the night before.

A short time later, sheriff's deputies identified the owner of the pager as Mark Bogan, who coincidentally owned a white pickup that had obvious damage to the rear end. Bogan immediately became the primary suspect. With the pager as damaging evidence, Bogan willfully admitted to picking up the hitchhiking Johnson and having sexual intercourse with her. However, he vehemently denied having anything to do with her homicide. He claimed that after having sex, the two began to argue, and when he demanded that she get out of his truck, she reached up onto the dashboard and stole a number of items from him, including his pager.[5] Bogan assured the investigating officers that he had never been in the same area as where the victim's body had been found.

Since Bogan had admitted to picking up the victim, the idea of linking Denise Johnson to the suspect's pickup was irrelevant. The investigators would have to link Bogan to the area in which the victim's body had been found. With Bogan's statement in mind, detectives searched the suspect's pickup for any evidence that could connect him to the crime scene — and what lay in the truck's bed was exactly what they needed. Detectives found pods from a plant or tree that they could not identify until returning to the scene where the victim was found. Back at the crime scene, they noticed that the paloverde trees that surrounded the area possessed the same pods that were recovered from the suspect's pickup. Theorizing that the pods found within the bed of Bogan's truck may have come from one of the gouged trees, the detectives questioned whether it was possible to link the two pieces of evidence through genetic DNA comparison.[6]

Following an extensive search for a genetic researcher who could accurately perform the necessary testing, Judge Susan Bolton assigned the complicated task to Dr. Timothy Helentjaris, a molecular geneticist from the University of Arizona. Using RAPD technology, Helentjaris not only profiled the pods collected from Bogan's truck, but also created a database comprised of the trees located nearest to the victim and others around the crime scene as well.

This small database, along with the fact that the paloverde tree population contains multiple genetic variations, allowed Helentjaris to conclude that the pods located within Bogan's truck bed most closely matched that from the tree nearest the body.[7]

However, since RAPD plant analysis had never been admitted into a criminal trial, a Frye hearing was necessary. At the hearing, Northern Arizona University professor Dr. Paul Keim testified on behalf of the defense and challenged the reliability and validity of not only the plant RAPD method but the sample collection and database creation as well. Dr. Keim insisted that in order to accurately determine whether the pods found in the suspect's truck were truly from a tree found at the crime scene, a much larger database was needed for statistical significance. Subsequently, Judge Bolton ruled that the evidence submitted by Dr. Helentjaris could be heard during the trial, but due to Dr. Keim's database challenges, no statistical random match probabilities could be used. With the aid of plant RAPD testing, good police work, and an abundance of other forensic evidence, Bogan was sentenced to life in prison, which was upheld by an appellate court in 1995.[4]

Another case involving the use of genetic plant typing in order to link a murder suspect to a crime scene occurred in Southern Finland in late 2001 (see Figure 15.1). A man, after meeting with his three criminal partners, was found dead about 5 km from where they were all last seen. Following the arrest of the three suspects, an analysis of their shoes, clothing, and car produced small pieces of botanical material known as bryophytes. These small samples of moss were the only pieces of evidence that detectives believed could link the suspects to the crime scene and the victim.

After visual analysis using a compound microscope, species identification was determined from samples found on one of the suspect's pants (*Calliergonella lindbergii*), from the suspects' car (*Brachythecium albicans*), and a suspect's shoe (*Ceratodon purpureus*). On returning to the crime scene, investigators collected multiple moss samples varying from 5 to 40 m from where the victim's body was found. These samples, used as known references to compare to the evidentiary submissions, were all later identified as *C. lindbergii*, *B. albicans*, and *C. purpureus*. After careful consideration of available genetic plant typing methods, investigators concluded that a PCR-based RAPD technique consisting of 10 arbitrary oligonucleotide primers and 17 simple sequence repeat primers (SSR) was to be used for the individualization of the plant samples.

Following the analyses of each evidentiary sample's genetic profile and those collected from the crime scene, statistical pairwise distances were compared in order to determine the genetic variation between each sample. Remarkably, it was concluded that the genetic analysis of the sample removed from the suspect's pants (*C. lindbergii*) might have originated from the crime

Figure 15.1 Bryophytes are commonly known as mosses, and their typical habitat is moist and shady as in wooded areas and along stream beds.

scene, and in particular, the area closest to where the victim's body was found. In addition, analysts found it to be likely that the sample collected from the suspect's car (*B. albicans*) also originated from the crime scene. However, due to the fact that *C. purpureus* commonly reproduces through sexual means (rather than clonal propagation like the first two plant species), considerable genetic variations made it too difficult to analyze using the selected DNA typing methods.[8] Ultimately, the results of the DNA typing analyses of these three plant species resulted in a failure to exclude these evidentiary samples from originating from the scene of the crime.

15.2.4 Restriction Fragment Length Polymorphism (RFLP) Analysis

Forensic restriction fragment length polymorphisms (RFLPs) are the variations in fragment lengths of DNA following digestion with one or more restriction endonucleases outside gene coding regions. The variation in the fragment length is dependent on where specific restriction endonuclease recognition sites are located throughout the species' genome. The restriction enzymes, used to cleave the DNA into identifiable fragments, cut at distinct 4 to 6 base pair sequences that are separated for analysis by gel electrophoresis.[9] The resulting fragments are compared to known size standards, and an estimation of fragment size is determined through visual comparison. The variation in fragment size from one sample to another is the result of base additions, deletions, substitutions, duplication, and translocations within the specific restriction enzyme recognition site.[10] The similarities in the fragment pattern can be used to aid in differentiating not only species from one another but strains as well. Although typically not used for forensic plant testing, RFLP has the capability of being a valuable tool in the field of forensic botany. RFLP analysis with plant samples would suffer the same constraints that have limited the use of RFLP technology in human forensic testing. These include the fact that a large quantity of DNA is required for testing since no PCR amplification step is performed and that the DNA must be of sufficient quality and purity for the test to yield a decipherable result.

15.2.5 Amplified Fragment Length Polymorphism (AFLP) Analysis

Amplified fragment length polymorphism (AFLP) analysis is an extremely powerful DNA fingerprinting method that offers more advantages than the precursor RAPD and RFLP techniques. Used for both the identification and DNA typing of eukaryotes and prokaryotes, AFLP analysis detects the presence of DNA restriction fragments by way of PCR amplification.[11] When AFLP is used for plant typing, DNA can be easily purified by the use of several commercially offered kits (e.g., DNEasy Plant System, QIAgen Inc., Valencia, CA). Following quantification by agarose gel electrophoresis, the isolated DNA is digested by two restriction endonucleases, cutting the DNA into many fragments.

Double-stranded adaptors are then ligated to each end of the restriction fragments, and preselective primers, with a single nucleotide bound to the

3' end, are used to perform preselective amplification specific for approximately 1/4 of the fragment pool. The second amplification stage, known as selective amplification, uses the product from the preselective PCR as a template for the selective primer binding. The selective primers are identical to the first set, with the addition of two more nucleotides downstream from the initial nucleotide on the 3' end of the primer. This allows for selective amplification of the restriction fragments to occur specifically for 1/16 of the previously amplified preselected product.[10] The selectively amplified product can then be resolved by acrylamide gel electrophoresis (see Figure 15.2).

15.2.6 Using AFLP to Track Illegal Marijuana Cultivation and Distribution

Within the field of forensic botany, the plant that has recently received the most attention in terms of research and development is marijuana (*Cannabis sativa*). Marijuana is a dioecious annual plant, which has been used by humans for more than 10,000 years. From the use of its seeds to produce oil and as a major grain source in ancient China to its leaves and stems to create strong rope and durable clothes, marijuana has been shown to have useful applications.[12,13] However, *C. sativa* is most commonly recognized as the plant that produces the illegal psychoactive drug Δ-9-tetrahydrocannabinol (THC).[14] For this reason, the Connecticut Department of Public Safety Forensic Science Laboratory is currently creating an AFLP marijuana database with the hope of linking distributors and users of the illegal narcotic to cultivators and their growing operations.[15] The concept for drug tracking is simple: many large-scale propagators have shifted from growing marijuana from seeds to the more productive and lucrative cloning method. This guarantees the cultivator that he will recreate the same genetic plant over a number of generations, and with those plants come the same, or in some cases even greater, THC content. Therefore, if a grower comes into possession of a very "potent" marijuana plant, he can generate large numbers of genetically identical plants.[16] By using the cloning method, an individual can begin to produce "bud" (which contains a majority of the psychoactive THC) in as few as eight weeks. This leads to a greater number of growing generations per year, which results in an increase of dividends.

15.2.7 Advantages and Disadvantages of AFLP Analysis

A major advantage of AFLP with respect to forensic botany is the ability to analyze multiple plant species and even individualize highly inbred plants. This is due to the vast number of fragments that are produced during the PCR amplification process. With a maximum number of 64 possible primer sets available, this enables one to create an extremely discriminating composite

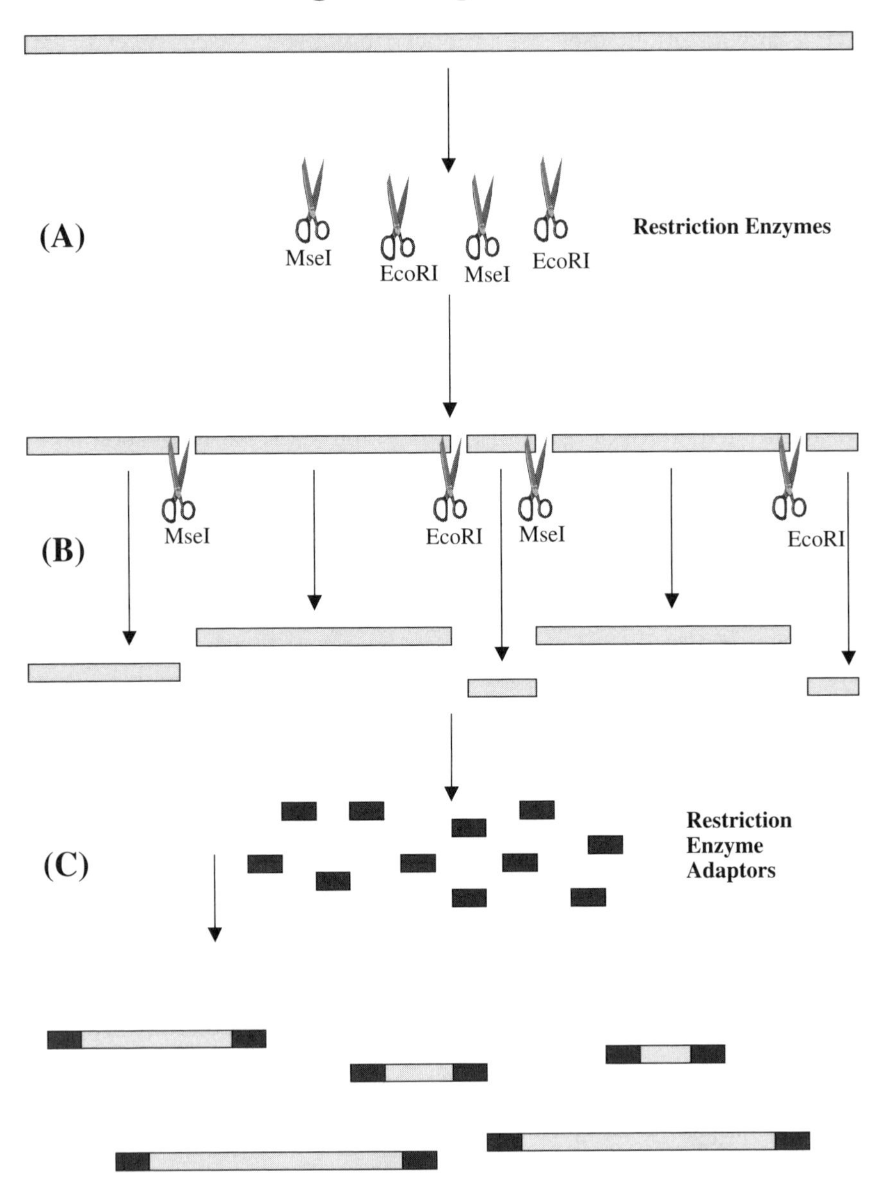

Figure 15.2 Amplified fragment length polymorphism (AFLP) analysis is essentially a combination of restriction fragment length polymorphism (RFLP) and polymerase chain reaction (PCR) as the general method involves the amplification of digested genomic DNA fragments to yield a band pattern that can be used as an identifier.

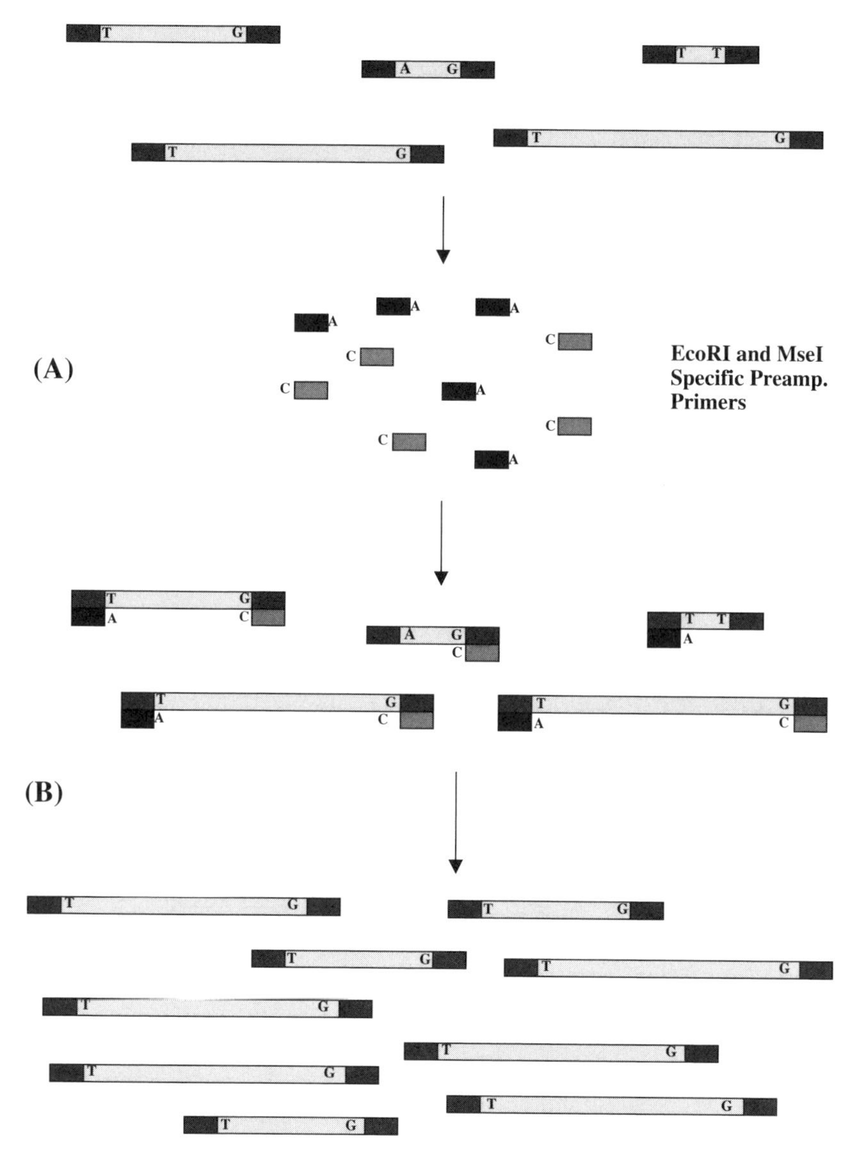

Figure 15.2 (continued)

DNA profile consisting of thousands of fragment peaks. Another advantage of the AFLP technique is that the genomic sequence of the plant sample is not required. This is important because DNA sequencing of an organism's genome is time consuming and expensive, and would need to be performed

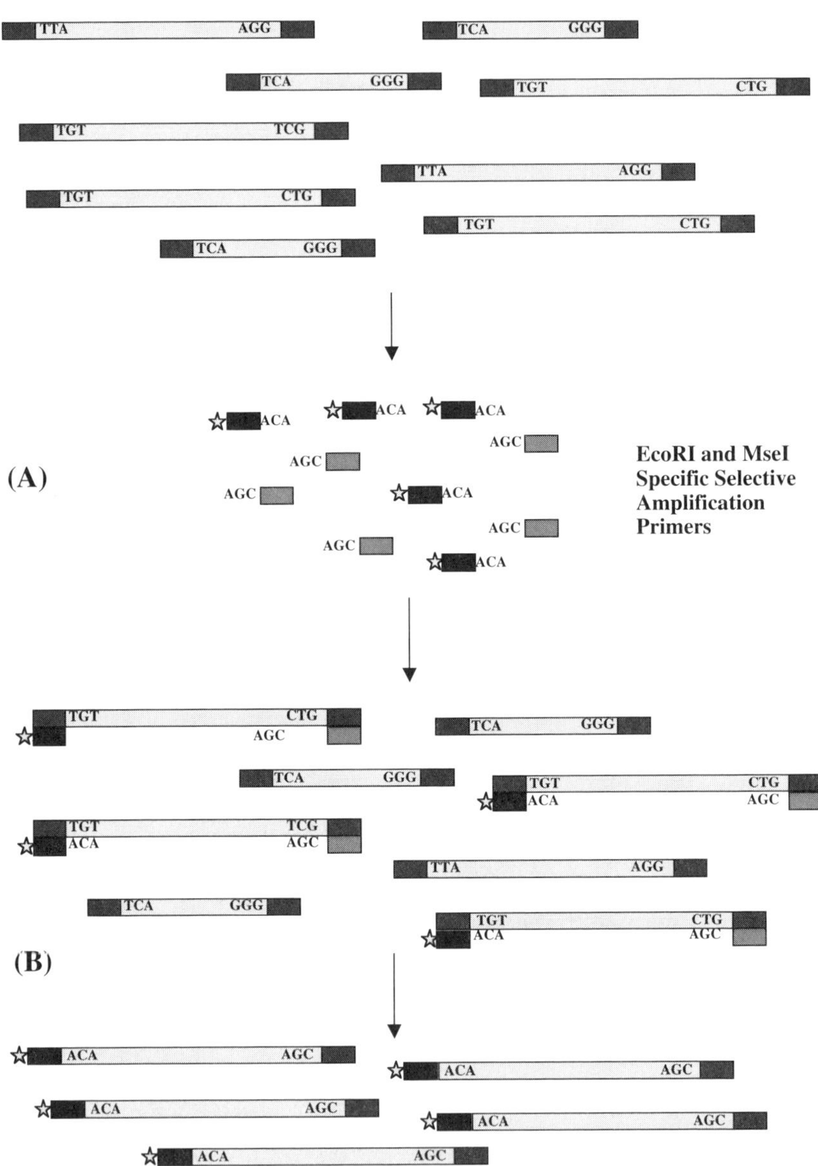

Figure 15.2 (continued)

for every plant species applied to casework. Since the DNA is being fragmented based on restriction enzyme sites, AFLP analysis can be applied to any plant species or biological organism under study. This means that the same AFLP technique that is used for tracking marijuana cultivars can be used to link an evidentiary plant sample of any species to a crime scene, a suspect, or victim.

A disadvantage of AFLP is the inability to differentiate between mixed and single-source samples. With the large number of fragments being created, knowledge of the plant sequence would be useful for determining which chromosome each fragment is located on to establish genetic linkage relationships. An understanding of the linkage relationship between fragments is necessary for calculating a highly discriminating statistical random match probability using the product rule method (allelic frequency at locus #1 x allelic frequency at locus #2, etc.) analogous to human STR profiling. However, without information pertaining to the genetic association between different AFLP fragments, the conservative counting method (number of times observed/total number in population) may be used to determine the forensic relevance of a sample DNA profile match.

15.3 The Future of Forensic Plant Genotyping

15.3.1 Repeat Sequence Genetic Testing

Within the field of forensic science are many methods of testing biological material that are based on defining the number of tandemly repeated nucleotide sequences of varying lengths. On average, plants contain a single short repeat sequence every 33 kilobases (Kb) of DNA as compared to the human average of one repeat unit every 6 Kb.[17] The number of nucleotides per repeat unit (also known as the core sequence) can range from two to several thousands of bases. Most commonly known as satellites, these repeated sequences are categorized into long-, medium-, and short-length repeats (forensic testing utilizes the latter two types of repeats). The medium-length repeats (minisatellites or variable number tandem repeats [VNTR]) are within the range of 10 to 100 bases in length while the short-length repeats (short tandem repeats [STR]), simple sequence repeats (SSR), or microsatellites range from 2 to 6 bases per repeat unit. However, the use of trinucleotide (3) or tetranucleotide (4) base repeats is most commonly used in human forensic testing.[18]

Within the past two years, STR markers for identifying clonally cultivated and seed propagated marijuana have been identified. Research scientists in Connecticut, Florida, and Taiwan are currently in the early stages of creating both mono and multiplex STR tests that may aid in the international surveillance and enforcement of marijuana propagation and the drug-trafficking trade. Researchers at Florida International University have discovered polymorphisms within 11 marijuana loci that have led to the identification of 52 alleles with frequencies ranging from 0.015 to 0.773. The testing combination of three dinucleotide and eight trinucleotide markers not only led to a random match probability of 1.8×10^{-7} with respect to a 41-sample database, but comparative AFLP testing, performed at the Connecticut State Forensic Science Laboratory, also demonstrated 100% concordance between the two (STR vs. AFLP) marker systems[16,19]. At the same time, researchers at the Central Police University in Taiwan have discovered a single, highly polymorphic marijuana STR marker that could contain 3 to 40 hexanucleotide repeat units. After processing 108 marijuana samples, this 6-base repeat demonstrated a high degree of genetic heterozygosity. Cross-reactivity studies performed with up to 20 different species (including the closely related *Humulus japonicus* and *Nicotiana tabacum* species) produced no amplified PCR product with either the forward or reverse primers.[20] Additional research studies have also led to the isolation of 10 more STR markers that were screened with 255 marijuana samples from over 30 countries.[16] Although further optimization and validation experiments need to be performed, current research trends indicate a major push toward producing a multiplexed STR marker system for marijuana. This system could provide valuable information to law enforcement authorities concerning the issues of genetic relatedness, use, propagation, and distribution of marijuana, not only across the U.S., but also throughout the world.

15.3.2 Advantages and Disadvantages of Short Tandem Repeat Analysis

Once STR markers are identified and developed in a species, one of the major advantages for using STR markers is the relatively short amount of time it takes to create a DNA profile as compared to other methods, such as AFLP and RAPD analyses. With AFLP analysis, the amplification of multiple primer sets could take several hours to set up, digest and ligate the DNA, undergo preselective and selective PCR amplification, and perform fragment detection. Some of the time required for AFLP is dependent on how many of the possible 64 combinations an analyst has chosen to use for selective PCR. For STR markers, the hands-on time is limited following the development and optimization of a single multiplex PCR reaction since each evidentiary and reference sample requires amplification once rather than multiple times. In addition, multiplexing a number of loci into one PCR reaction increases high-throughput processing capabilities, and saves time compared with processing multiple monoplexes.

Another advantage of STR markers is the ability to produce full DNA profiles from low DNA quantity or degraded samples. With AFLP and RAPD analyses requiring multiple nanograms (10 to 20) and high molecular weight DNA (typically >6Kb), preliminary validation experiments with degraded and low-level marijuana samples have demonstrated that STR primers have the ability to effectively amplify samples below 6 Kb and at 0.5–10 ng quantities.

Although further validation is required before making definitive conclusions, as with human minisatellites, plant STR markers appear to work well with samples of low genomic quantity.

The third advantage to plant STR markers is the ability to use more discriminating statistics when calculating the random match probability of an evidentiary sample. When sequencing the repeat motifs, identifying where and on which chromosome a marker maps to is important for showing that the loci can be inherited independently (nonlinkage). This genetic information defines which statistical method to use. As demonstrated with the set of marijuana STR markers from the Florida International University research, the probability of nonclonal common genotypes was estimated in the range of 1 in over 5.5 million plants, but linkage studies need to be performed.[19]

Plant STR markers are also comparatively easy to analyze. Unlike AFLP analysis, which requires converting a large subset of fragments to binary code for comparison of sample profiles, STR markers are scored by comparing fragment size to an allelic ladder (fragments of known size and repeat units). Computer software is then used to convert the fragment sizes to allelic values. Both AFLP and STR analyses generate a string of numbers to represent a DNA profile. However, due to the complexity and greater number of fragments generated by AFLP, the analyst still needs to compare the final AFLP profile by overlaying the peaks since not all fragments generated are scored as a binary code. With STR markers, the amplification process produces a low number of peaks per locus (usually two), which accounts for easier comparison between known and evidentiary samples.

One final advantage to plant STR markers is their ability to differentiate between mixtures. Normally known as a 20-chromosome (diploid) plant, marijuana mixtures can be determined based on the number of observed peak fragments at a given locus. A normal diploid cell contains a maximum of two peaks (alleles) per locus; one from the pistillate (mother) plant and one from the staminate (father) plant. When three or more peaks occur at multiple loci, a mixture of two or more contributors is most likely present. Analysts can then use their training and expertise in order to determine the inclusion or exclusion of the evidentiary profile as compared to the known reference sample. For those labs using AFLP analysis as their major plant typing technique, STR markers can be used as a preliminary method for determining the number of DNA contributors to a sample. However, this can only be accomplished if the STR marker sequence for the plant species is known.

The first major limitation of plant STR typing is that the sequence of the microsatallite and the flanking region must be known. To accomplish this task takes many man-hours of sequencing, money, and the knowledge of molecular and cell biology, genetic engineering, and applied plant genomics. Therefore, it would be impractical to create a new STR test for more than a few plant species because sequence information and a species population database would need to be available. Considerable sequence information is essential for creating a multiplex PCR assay, and the individual loci must demonstrate linkage equilibrium (independence) in order to calculate the individual match probabilities.

Another disadvantage to STR plant typing is the difficulty in optimizing the reaction conditions for the multiplex. One benefit of STR markers is that they can be amplified in a single reaction step. However, in order for this to occur, a multiplex reaction, consisting of primers that are specific for different loci, must be created and optimized. Performing this task may be challenging due to the difference in annealing temperatures, $MgCl_2$ concentrations, primer quantities, and fluorescence intensities of the different 5'-modifications between all forward and reverse PCR primer pairs. Following optimization, interlaboratory validation and population databases must be created for each plant species with the corresponding multiplex. With this in mind, many research laboratories will choose to create STR multiplex kits for plant species commonly encountered throughout the U.S., such as marijuana STRs.

15.3.3 Single Nucleotide Polymorphism Assays

Single nucleotide polymophisms (SNPs, snips, or singletons) are point mutations in which a single nucleotide is substituted for another at a certain locus.[21] Known as the most common type of genetic variation in both plants and animals, within the field of forensic botany an abundance of these single polymorphisms can be used to differentiate an evidentiary sample from a known reference submission. With the popularity of SNPs on the rise, many companies and research facilities have published a variety of techniques on how to analyze and map SNPs using highly technical methods. These innovative techniques demonstrate the accelerated rate at which plant DNA typing technology is advancing.

One of the more popular methods for detecting the presence of a SNP is that of the Affymetrix chip technology. This involves the use of a microchip embedded with thousands of oligonucleotide probes. When fluorescently labeled amplified DNA is bound to the chip, it is allowed to hybridize to the probe, producing a luminescent pattern that indicates the presence of a single nucleotide polymorphism.[18,22] With the potential for a single chip to detect thousands of SNPs, it is clear that these markers can be a powerful tool for analyzing a variety of plant samples and species. Although not currently used for forensic plant genetic typing, the rate at which SNP technology and research is advancing predicts the possibility of using forensic plant SNPs in court in the not-too-distant future.

15.3.4 Advantages and Disadvantages of Plant SNP Assays

Compared to plant STR markers, SNP assays are less informative per marker yet possess many advantages that could make them a forensic plant profiling technique in the future. One of the major advantages already mentioned is the ability to detect thousands of SNPs on a single microchip. Although this may appear to be an incredibly high number, within years, microchip technology will most likely allow for millions of single nucleotide polymorphisms to be analyzed in half the time on a microchip half the size. Another advantage of SNP analysis is the fact that these markers are small and possess no PCR artifacts, such as stutter or minus A, and thus are exceptionally useful for low-level degraded samples.[18]

An obvious problem that is encountered with plant SNPs is the fact that much sequencing and a large database would have to be created for every plant species tested. Unlike the human-derived SNP database (SNP Consortium Ltd.), which can be applied uniformly to all *Homo sapiens*, every plant species is different. Another disadvantage of SNPs is that, for a STR marker that has 20 or more alleles, anywhere between 30 to 50 SNPs will have to be developed to obtain equal frequencies and discrimination power.[23] In addition, the price of manufacturing a single microchip can be costly because of the expertise and technologically advanced equipment required to create them properly.

15.3.5 Microchip Nanotechnologies

With the application of new plant typing techniques (e.g.,. STRs and SNPs) comes novel methods of analyzing them. Recently, and more so in the near future, the validation of high-throughput SNP technology will be essential for widespread forensic application. As previously mentioned, Affymetrix's microchip has allowed for allelic detection in a fraction of the time, with less hands-on bench work and a large number of SNP detections. However, Affymetrix is only one of many companies currently in the process of manufacturing technologically advanced genetic-based microchips for possible future forensic applications. Although aimed at human allele identification, the use of this technology in plant DNA typing is technically feasible and could be realized within the next few years.

Another popular company in the midst of microchip technology is Nanogen Inc. (San Diego, CA). Most promising for the possible future use with specific plant STRs, Nanogen has created a silicon chip (NanoChip® Electronic Microarray) with multiple, individualized test sites that have the capability to bind precise repeat sequence alleles. Measuring one square centimeter, the nanochip contains 100 test sites that are approximately one square millimeter and are controlled by individual electric potentials.[24] Specific capture probes, containing the complementary sequence of each set of repeated alleles in a locus, is bound to each site, along with a fluorescent reporter probe for detection purposes.

A DNA sample is washed over the chip and binding occurs only when an exact match of an STR allele hybridizes to a matching capture probe. Therefore, an eight-repeat tetranucleotide STR will only bind to the test site on the microchip that contains the complementary, eight-repeat tetranucleotide capture probe for a particular locus. When this occurs, the reporter probe fluoresces and is detected by state-of-the-art software. When each STR has bound to its specific test sites, the genetic profile of the sample is created by which test sites have fluoresced and which have not.[18,23] Although not designed for forensic plant DNA genotyping, the capability of the NanoChip® Electronic Microarray to produce botanical STR profiles is very likely.

Another advance in microchip technology is the capillary electrophoretic (CE) microchip that has the potential to accommodate high-throughput sampling with ease. First created in 1992, the CE microchip's ability to separate fluorescent dyes, restriction fragments, STRs, and SNPs makes it a logical technology for future plant genotyping. The typical design of the CE microchip entails a glass slide with etched grooves, covered with another glass slide in order to create individualized channels. Up to 100 times faster than standard electrophoresis, CE microchips have to ability to separate STR markers in a matter of seconds rather than minutes or hours. In addition to CE microchips, capillary microplates have been designed, which enables the separation of 96 different samples within 2 minutes.[23,25]

15.4 Plant DNA Extraction: Combating the Problem of Low Yield

One of the more difficult aspects of working with botanical evidentiary samples is the issue of limited DNA recovery using traditional plant DNA extraction methods. The common method of soaking the sample in liquid nitrogen and then grinding it into a powder by hand is both strenuous and time consuming. In addition, as samples are processed, the ones that were previously pulverized need to be kept cold, so the analyst must repeatedly add liquid nitrogen to the rack in which they sit in order to ensure that samples do not thaw. A batch of 20 samples could take up to or over an hour to grind, which often results in low DNA recovery, and limits the number of samples that can be processed at any one time.

Qiagen Inc. (Valencia, CA) and Retsch Technologies GmbH (Haan, Germany) have worked together to combat this problem with the Mixer Mill MM300. Able to pulverize multiple samples at once, this machine enables high throughput and effective homogenizing of 192 plant samples (two 96-sample racks) in less than 3 minutes. Frozen plant material, along with a stainless steel

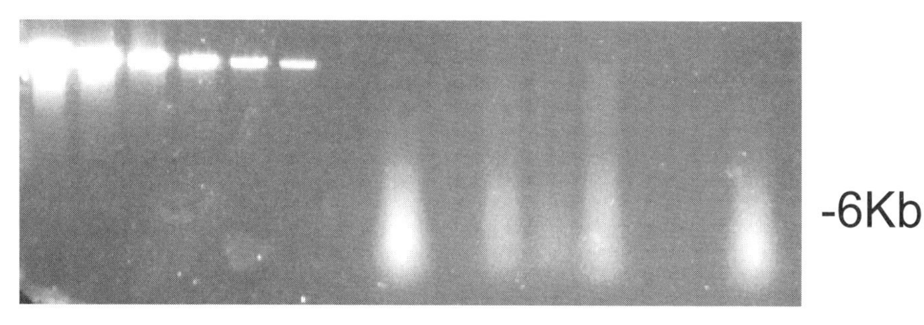

Figure 15.3 An example of plant DNA recovered from ten-year-old dried plant leaf samples. Although the DNA is significantly degraded, it still met validation criteria and a valid AFLP profile was generated from each sample. Lanes #1–6 are yield standards of 400, 200, 100, 50, 25, and 12.5 ng respectively; lane #7 is a reagent blank control; lanes #8–15 are various samples from the dried *Jacaranda* leaf material.

or carbide bead, is added to pre-racked tubes and is inserted into the MM300 and pulverized through high-pulse shaking (up to 30Hz). This vibration causes the bead to mechanically grind the plant sample into a fine powder that allows for a greater degree of cell disruption, as compared to manually ground samples. Qiagen has shown that following chemical extraction, the samples that were purified using the MM300 resulted in greater yield.

With respect to forensic botany, the Mixer Mill 300 may be the answer to the problem of DNA recovery from low-level samples. Typically, the amount of DNA needed to type a plant evidentiary sample is 20 or more milligrams. However, evidentiary samples are not always of sufficient size to yield the required quantity of DNA. Whether it's a marijuana sample found on a victim that is being compared to a suspect's "stash" or a piece of leaf found in the cuff of a suspect that may link him or her to a crime scene, the quantity of plant material may be marginal. With the help of mechanical-based cell disruption, the amount required may be reduced substantially. Preliminary validation performed at the Connecticut Forensic Science Laboratory has shown the ability of the Mixer Mill 300 to help extract DNA from 10-year-old, room temperature-stored, low-quantity dried plant material (~15 mg) (see Figure 15.3).

15.5 Field Testing: Genotyping at the Crime Scene

As described earlier in the paloverde and bryophyte cases, the collection of botanical evidence at a crime scene can play an important role in the investigation process. At a crime scene the amount of evidence that is collected may be very large, and a major problem to investigators is differentiating between which samples are important and which are not. At an indoor crime scene, the issue of which plant samples to collect may be limited. However, when the crime (such as homicide or sexual assault) occurs in a wooded area, the matter of botanical evidence plays a much more significant role. Normally, if a body is found within a wooded area, detectives are forced to spend many hours processing the crime scene and collecting plant evidence anywhere from inches to many meters away from the victim, but which pieces of evidence are truly worth collecting? If the samples could be processed and analyzed minutes after being collected, this may aid not only the investigation but reduce the pressure for the laboratory analysts to produce results and conclusions for such a large number of samples in such a short period of time.

In the near future, crime scene investigators may be able to process hundreds of samples in mere minutes. This notion of crime scene analysis began to become a reality in 1998 when the Lawrence Livermore National Laboratory engineered a miniature analytical thermal cycler instrument (MATCI). Transported within a suitcase and weighing 35 pounds, this miniature marvel consisted of a Macintosh PowerPC laptop and a microfabrication instrument. Powered by 13 nickel–cadmium batteries, completing 30 cycles in less than 45 min, and having the capability of real-time multifluorescent detection, the MATCI enabled forensic analysts to determine whether they had enough amplified product from individual samples.[26] The capability of combining portable PCR with high-throughput microchip CE technology may allow a crime scene investigator to process multiple samples in a mere 20 minutes. In 1998 there was no reliable, portable method in which to genotype the amplified samples in the field, but the capacity to do this may be a reality in the near future.

However, in order to process evidentiary samples directly at a crime scene certain obstacles must first be overcome. Primarily, by performing the amplification procedure on scene, the chances of contaminating a sample with foreign DNA is greatly increased. Since PCR is extremely sensitive, certain precautions (e.g., use of laminar flow hoods) must be taken to ensure that extraneous DNA is not incorporated into the sample, and as a result the eventual profile. Besides the issue of contamination, it is widely known that in order to process forensic evidentiary samples, a laboratory must be acknowledged as an accredited forensic facility by ASCLD-LAB (American Society of Crime Lab Directors–Laboratory Accreditation Board). With strict guidelines pertaining to sample handling and packaging, operating procedure, and testing facilities, it may be difficult to defend the notion that a crime scene is an adequate location for DNA processing. Another aspect that must be overcome before evidence can be processed at a crime scene is that generally it is police officers, rather than forensic scientists, who report to and collect evidence from a crime scene. With the increase in cases being submitted to forensic labs around the country, it may be difficult to have trained scientists report to crime scenes when they could process the samples "in house" to avoid a possible backlog that is being created in their absence.

15.6 Summary

After plant species identification has been performed using traditional microscopic methods or molecular DNA sequencing of specific genes, the determination of whether or not a plant sample has come from a specific plant (individualization) or group of clonal plants may be required. The use of plant DNA for the individualization of trace botanical evidence in forensic science is still in its infancy, but as science and technology advances, the challenges to these forms of DNA profiling will be resolved. With these scientific advances, one can predict a bright future for forensic botany and plant DNA typing and an increase in their use to aid in police investigations and criminal case resolution.

References

1. Williams, J.K., et al., DNA polymorphisms amplified by arbitrary primers are useful as genetic markers, *Nucleic Acid Research*, 18, 6531, 1990.
2. http://www.amershambiosciences.com.
3. http://cgn.wageningen.
4. Inman, K. and Rudin, N., *An Introduction to Forensic DNA Analysis*, CRC Press, Boca Raton, FL, 1997.
5. Day, A., Nonhuman DNA testing increases DNA's power to identify and convict criminals, *Silent Witness*, 6, 2001.
6. "Science America Frontiers," WNED-TV (PBS-Buffalo, NY), January 19, 1994.
7. Yoon, C.K., Botanical witness for the prosecution, *Science*, 260, 894, 1993.
8. Korpelainen, H. and Virtanen, V., DNA fingerprinting of mosses, *J. Forensic Sci.*, 48, 804, 2003.
9. Glick, B. and Pasternak, J., *Molecular Biotechnology: Principles and Applications of Recombinant DNA*, 2nd ed., American Society of Microbiology (ASM) Press, Washington, D.C., 1998.
10. AFLP™ Plant Mapping Protocol, Applied Biosystems, Inc., Foster City, CA, 2000.
11. Lin, J.J., Kuo, J., and Qiu, J., Applications of AFLP markers in prokaryotes and eukaryotes: from radioactive detection to nonradioactive detection, presented at Plant & Animal Genome VII Conference, January, 1999.

12. Siniscalco Gigliano, G., Preliminary data on the usefulness of internal transcribed spacer I (ITS1) sequence in *Cannabis sativa* L. identification, *J. Forensic Sci.*, 44, 475, 1999.
13. Dreyer, J. et al., Comparison of enzymatically separated hemp and nettle fibre to chemically separated and steam exploded hemp fibre, *J. Industrial Hemp*, 7, 43, 2002.
14. Clarke, R.C., *Marijuana Botany-An Advanced Study: The Propagation and Breeding of Distinctive Cannabis*, RONIN Publishing, Inc., Berkeley, CA,1981.
15. Silverman, L., *Pot Prints: Making a Database of Marijuana DNA*, see Court TV.com, July 25, 2003.
16. Miller Coyle, H. et al., An overview of DNA methods for the identification and individualization of marijuana, *Croatian Medical Journal*, 44, 250, 2003.
17. Wang, Z. et al., Survey of plant short tandem DNA repeats, *Theor. Appl. Genet.*, 88, 1, 1994.
18. Butler, J., *Forensic DNA Typing*, Academic Press, San Diego, CA, 2001.
19. Alghanim, H. and Almirall, J., Development of microsatellite markers in *Cannabis sativa* for DNA typing and genetic relatedness analyses, in press, 2003.
20. Hsieh, H-M. et al., A highly polymorphic STR locus in *Cannabis sativa*, *Forensic Sci. Int.*, 131, 53, 2003.
21. Rafalski, A., Application of single nucleotide polymorphisms in crop genetics, see www.esb.utexas.edu.
22. Wang, D.G. et al., Large-scale identification, mapping, and genotyping of single-nucleotide polymorphisms in the human genome, *Science*, 280, 1077, 1998.
23. Stuart, J.H. and Jon, N.J., *Forensic Science: An Introduction to Scientific and Investigative Techniques*, CRC Press, Boca Raton, FL, 2003.
24. http://www.nanogen.com.
25. Simpson, P.C. et al., High-throughput genetic analysis using microfabricated 96-sample capillary array electrophoresis microplates, *PNAS*, 95, 2256, 1998.
26. Belgrader, P. et al., Rapid PCR for identity testing using a battery-powered miniature thermal cycler, *J. Forensic Sci.*, 43, 315, 1998.

Appendix A: Considerations for the Use of Forensic Botanical Evidence: An Overview

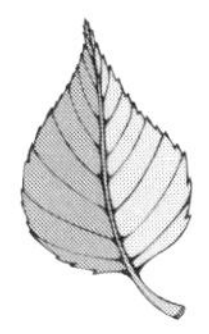

When considering the use of forensic botanical evidence found at a crime scene, one must eventually factor in the integration of crime scene collection, forensic laboratory testing, and finally, the presentation of the evidence in court. The following questions and suggestions are designed as a guideline to assist practitioners at each step of the integrated process.

A.1 At the Crime Scene

- A review of the entire crime scene should be made to determine if any botanical evidence might have been overlooked for collection. In particular, small plant fragments, seeds, pollen, and flowering plants should be noted.
- For bodies, have all of the seams, pockets, cuffs, shoes, etc. of clothing been checked for adherent plant matter? Have all of the body orifices been examined for plant matter (e.g., algae, stems, seeds)?
- If plant matter seems to have a significant value at a crime scene, has the scene been reviewed to determine and document the location of any similar plant species (e.g., is there more than one plant representing the species at the crime scene, such as three holly bushes)? Reference samples need to be collected from all three for later examination.

0-8493-1529-8/05/$0.00+$1.50

- For vehicles, have the undercarriage, windshield wipers, vents, floor mats, and so forth been checked for seeds or leaf and flower fragments?
- Has a schematic diagram for indoor and outdoor aspects of the scene been made? All plants of significance should be noted.
- Sometimes through the use of plant ecology (knowledge of plant growth patterns and geography), estimation for time of death can be established. For example, if a plant is growing through skeletal remains, the age of the plant can be approximately determined and perhaps a season (or other time estimate) assigned for when the body may have been laid at that location.
- Have any flowering plants that may be pollen reference samples been noted? Samples should be collected for later species identification. Leaf and flower samples should be pressed flat and stored in paper envelopes to maintain original shape.
- If evidence has been identified, has it been packaged appropriately (e.g., paper)? If any unusual forms (pollen, fecal matter, etc.) are present, has a call been made to the forensic laboratory regarding packaging and storage temperature? This step may be critical if DNA identification and individualization steps are to be performed later at the laboratory.
- Have appropriate chain-of-custody records been maintained? Is all of the evidence sealed and initialed?
- Are any of the plant samples unusual in appearance (i.e., not a common species)? The less common a plant is, the higher the probative value for the case.
- Is there other forensic evidence in the case? Often, botanical evidence is not considered until the other forms of forensic evidence are deemed nonprobative. Consider in advance the fact that bloodstains may seem informative at the scene; however, if DNA testing later shows that they are all originating from the victim, then the additional evidence may become more significant to the case.

A.2 At the Forensic Testing Laboratory

- Check the chain of custody and storage conditions prior to the evidence's submission to the laboratory, especially if DNA testing may be performed later.
- Make a visual examination of the evidence for gross physical characteristics that may be useful for botanical identification to the species level:

- Leaf morphology — shape, size, texture, edges, color.
- Vein patterns in the leaf — where are the major veins and how do they connect to the minor veins?
- Flower morphology — shape, size, texture, edges, color, fragrance.
- Arrangement of flower on a stem or in relation to each other if multiple flowers are present.
- Seed morphology — shape, number, size, patterns or etching on the surface, color.
- Roots — texture (fine versus coarse), fibrous versus tap, size, color

- Perform a microscopic examination of the evidence for minute characteristics that may be useful for botanical identification to the species level:
  - Leaf morphology — surface features (e.g., trichomes, stomatal patterns); in a cross section, what is the arrangement of cell types (e.g., parenchyma cells versus epidermal cells)? Note presence or absence of inclusion bodies (e.g., crystals and their shapes/sizes).
  - Flower morphology — petal features, presence of fluorescence under ultraviolet light; in a cross section, is there any special arrangement of cell types? What is the shape and orientation of pigment containing cells?
  - Arrangement of tissue types and cells in a cross section for stems and roots can significantly aid in the identification of certain plant fragments.
  - Seed morphology — surface patterns that may be helpful in plant identification to the genus or species level.

- If plant species identification is in question after examining physical traits, DNA extraction may be necessary for the molecular identification of a plant species.
- Is the plant sample free of any visible mold or fungi or adherent pollen? Have all seeds been removed from the vegetative tissue? Is the plant sample of sufficient size to yield enough DNA for testing?
- Once DNA has been extracted, is the quality and quantity sufficient for further analyses?
- If species identification is to be performed using DNA, what types of comparisons can be made and are any known reference samples available?
- If the individualization of a plant species is to be performed using DNA, what will be the possible outcomes of the testing: a match, a no match, or inconclusive?

- If a match has been made between an evidentiary sample and a known reference sample, what is the probative value? For plant species that are "unusual" or rare, a match may be very significant unto itself. However, if a plant species is more common, a comparative reference database may have to be constructed to establish a statistical value for the match.
- Have the report conclusions been clearly stated and explained to the submitting agency?

A.3 In the Courts

- A review of the chain of custody and documentation should be made.
- A review of the report and the report conclusions should be made.
- Has plant evidence ever been introduced into the courts in your state?
- What are the qualifications of the examiners and/or those performing testimony?
- What are the qualifications of the laboratory performing the testing?
- If the examination and testing was performed at an academic institution, are there records of validation for the use of the techniques and the experience of the examiner with court testimony? If not, a careful review may be necessary to avoid having the evidence not introduced into court.
- Was a blind test performed or possible when attempting plant species identification? This step may not be necessary with sufficient examiner experience and if adequate reference samples are available for comparison.
- Is the comparative database representative of the plant population that might be found at the crime scene? For example, plants selected from the local geographic region may represent the natural variation found within that plant species as opposed to those selected from archival samples in Herbaria from across the world. Both types of samples can provide useful information on levels of genetic variation in a plant species; however, some may be more relevant to the case under consideration. Herbarium samples are useful when a plant is rare and does not have a natural population from which to collect samples.
- What is the predicted level of discrimination power for the form of plant DNA testing used in the case? This may vary depending on the test method or the propagation method for the plant species in question. For example, in the case of clonally propagated marijuana,

linkages may be made between users, dealers, and growers based on the commonality of plant DNA profiles. The establishment of one hundred plants all with identical DNA in a localized geographic region strongly supports clonal propagation for sale and distribution. Alternatively, seed-propagated plant samples that match are especially probative because it is not anticipated that populations of identical plants are present in a limited geographic region.

- Are all of the participants in the case apprised of the benefits and limitations of the botanical evidence to avoid any discrepancies in testimony?

If these questions are asked prior to the use of botanical evidence in testing and in court, the evidence can be assured of the appropriate level of representation in criminal cases. Botanical evidence is inherently variable due to the great number of plant species present in land and water. A comprehensive understanding of the plant species in question, the test methods, and a cooperative effort between crime scene personnel, forensic scientists, academic practitioners, and attorneys can lend clarity to a judge and jury when forensic botany is presented in court.

Appendix B: Glossary of Terms

Acclimation Gradual physical and biochemical change that occurs in a plant to prepare it for winter conditions.

Active transport Transport of a substance across a cell membrane; a process that requires energy.

Adventitous Development of buds or roots from sites other than typical locations in the plant.

Aerobic respiration Respiration that occurs in the mitochondria and requires free oxygen.

Aleurone Outermost cell layer of the endosperm; high in protein and enzymes for the developing seed.

Alkaloids Toxic organic bases found in many plants; colorless, complex structure, and bitter.

Anaphase Stage in meiosis or mitosis where chromatids of each chromosome separate and move to opposite poles.

Angiosperm Flowering plants having seed-enclosed fruit.

Annual Plant that completes its life cycle in 1 year.

Apical dominance Suppression of lateral buds by hormones.

Apical meristem Group of actively dividing cells at the tips of roots and stems.

Apomixis Development of embryo from diploid parent tissue; asexual reproduction without fertilization.

0-8493-1529-8/05/$0.00+$1.50

Archegonium Female sex organ containing the egg in bryophytes, lower vascular plants, and gymnosperms.

Ascus Saclike reproductive structure of ascomycete fungi.

Atropine Alkaloid found in Belladonna.

Autotroph Organism capable of producing its own food.

Axil Site at a stem–leaf junction where a bud develops.

Basidium Clublike reproductive structure of basidiomycete fungi.

Biennial Plant that completes its life cycle in 2 years.

Biome Major regional ecosystem characterized by climate and having distinct dominant plant and animal species.

Bordered pit Pit in the walls of tracheids; overhanging secondary walls form a chamber between opening and pit membrane.

Bud Growing point of a plant shoot consisting of apical meristem and derivatives.

Bundle sheath Specialized layer of cells surrounding a vascular bundle.

Calcium oxalate Salt of oxalic acid making up crystalline deposits (druses or raphides).

Casparian strip In some plant roots, a specialized band of lignin and suberin encircling the epidermal cells.

Cell theory The theory that all cells come from preexisting cells.

Cereal Plant that produces grain or caryopsis.

Chlorenchyma Parenchyma cells that contain chloroplasts.

Chlorophyll A group of green pigments in plastids required for photosynthesis.

Chloroplast Cellular organelle with chlorophyll responsible for photosynthesis in eukaryotes.

Chromoplast Cell organelle containing carotenoids that give yellow, orange, and red color to plant parts.

Chromosomes Genetic structures in the nucleus composed of DNA and protein; allowing for inheritance of traits.

Clone Population of plants with identical DNA that are produced asexually from one parent.

Collenchyma Elongated living cell with irregular thickness to the primary walls; functions in support for the plant body.

Community In an ecosystem, all the plants represented.

Competition Two or more organisms competing for light, energy, space, or other resources.

Conifer Cone-bearing plant or gymnosperm.

Cork A protective tissue of dead cells in stems and roots.

Cork cambium Meristematic layer of cells that produces the cork.

Cortex A primary tissue composed of parenchyma cells; in stem and root storage tissues.

Cuticle A protective layer of cutin on the outer surface of the epidermis.

Cutin A fat substance deposited on outer epidermis walls of stems and leaves.

Cutting Typically a stem section used for vegetative propagation.

Cytoplasm The living protoplasm (jelly-like substance) of the cell excluding the nucleus.

Deoxyribonucleic acid (DNA) Nucleic acid that carries the genetic information from one generation to another.

Diatom Small single-celled photosynthetic organisms with silica-based walls (algae or protists).

Dicotyledon (dicot) A major angiosperm subgroup based on the embryo having two first leaves after germination of the seed.

Differentiation The further specialization of a cell as it matures.

Diploid A genetic condition of pairs of homologous chromosomes in a cell; 2N.

Dominant gene A gene that exerts a phenotypic effect to the exclusion of another allele.

Dominant species A plant species that predominates in an ecosystem based on size or numbers.

Ecology The study of organisms and their environment; interactions.

Ecosystem The group of organisms interacting with the environmental factors of the area.

Endodermis A special layer of cells in the root that helps regulate water passage.

Endosperm Triploid (3N) tissue in an angiosperm seed that provides nutrients for the developing embryo.

Enucleate A cell that no longer retains a nucleus.

Environment All the living and nonliving factors found in a specific region.

Epidermis The outer layer of cells of the primary plant body.

Ergastic substance A substance within the cell such as a crystal or inclusion body.

Etioplasts Plastids that have developed without the presence of light.

Eukaryotes Cells that have a membrane-bound nucleus and possess other membrane-bound organelles (plastids and mitochondria).

Fiber Sclerenchyma cell with thick secondary wall and functions in plant support.

Fibrous root system Roots with equal size fine roots interconnected; common in grass species.

Flower Reproductive structure common to angiosperms.

Genetic pool All the genes available to a population of breeding organisms.

Ground meristem Meristem tissue that develops into cortex and pith.

Guard cells Special epidermal cells surrounding an opening that is opened and closed by changes in turgor pressure.

Gymnosperm Plants that produce a seed that is not surrounded by a fruit structure.

Haploid A cell or organism having a single set of chromosomes (1N).

Heterotroph An organism that cannot produce its own food.

Hydrophyte Plant adapted to growing in water.

Hydroponics Plants grown in a hydroponic nutrient solution with or without a support medium.

Hyphae Thread-like structures that comprise the body of a fungus.

Hypocotyl The part of the embryo between the root and the cotyledons.

Hypothesis A working explanation that must be tested by experimentation to prove.

Inflorescence A floral cluster.

Intercellular spaces The spaces found between adjacent cells, e.g., parenchyma cells.

Interfascicular cambium The vascular cambium found between the vascular bundles.

Interphase A stage between mitotic cycles.

Intracellular Found within the cell.

Juvenile phase A rapid period of growth typically early in life.

Krebs cycle Also called the citric acid cycle; the second phase of aerobic respiration.

Lamina The leaf blade; flat portion of the leaf.

Leaflets The smaller parts of a compound leaf.

Lenticel A specialized form of cork for gas exchange found on young woody stems.

Lignin A compound found in secondary walls giving them strength and durability.

Limiting factor An environmental factor that limits growth of an individual organism or population.

Lipid Organic compounds such as fats, oils, and waxes.

Mature phase An adult growth period following the juvenile phase when sexual reproduction may occur.

Megaspore A spore that germinates to produce a female gametophyte.

Meiosis Two nuclear divisions where chromosomes are reduced from 2N to 1N and the chromosomes are distributed in 4 (1N) cells.

Mesophyll The area between the upper and lower epidermis that is photosynthetic.

Mesophyte A plant adapted for growth in an area of high moisture content.

Metabolism The sum of all chemical reactions within a living organism.

Metaphase A stage in meiosis and mitosis where chromosomes are arranged at the equator of the cell.

Microtubules Hollow rod structures within a cell that participate in nuclear division and cell movements.

Middle lamella A pectin "glue" layer that holds the primary cell walls together.

Mitochondrion Plural; mitochondria, a cellular organelle for respiration in eukaryotes.

Mitosis Division of the nucleus to produce two nuclei that have the same chromosome set as the parent cell.

Monocotyledon In flowering plants, a plant with one cotyledon leaf, e.g., grasses, orchids.

Mutation Heritable change in a gene or chromosome.

Mutualism A symbiotic relationship beneficial to each organism.

Mycelium A mass of hyphae.

Netted venation Veins are arranged in an interlocking net pattern.

Node The part of a stem where a leaf is attached.

Nonvascular plant "Lower" plants lacking xylem and phloem.

Nuclear envelope The double membrane layer that surrounds the nucleus in eukaryotes.

Nuclear organizer Area on a chromosome where nucleoli are assembled.

Nucleolus Dense bodies in the nucleus; ribosome precursors.

Nucleotide Structural component of DNA and RNA.

Nucleus The largest cellular organelle in eukaryotes; contains the chromosomes.

Omnivore An organism that eats both plants and animals for nutrients.

Organ A plant structure composed of multiple tissues.

Organelle A specialized structure within the cell with a specific function.

Ovule An organ located within the ovary of a seed plant where fertilization occurs.

Palisade parenchyma Photosynthetic cells comprising the upper layer of mesophyll cells in leaves.

Parallel venation A parallel arrangement of the main veins in a leaf; typically monocots.

Pectin A binding agent in the middle of the primary plant cell walls.

Pericycle The layer of cells inside the endodermis from where lateral roots develop.

Petiole The leaf stalk.

pH Indicates the hydrogen ion concentration in solution; can be basic, acidic, or neutral.

Phloem Conducting tissue for nutrients.

Photorespiration C3 plant process that reduces carbohydrates obtained by CO_2 fixation.

Photosynthesis The conversion of sunlight into chemical energy.

Phylogeny The evolutionary history for a group of organisms.

Phytochrome Plant pigments associated with far-red and red light.

Phytotoxins Plant poisons.

Pioneer plant A plant that colonizes a previously uninhabited site.

Pistil The term for a female reproductive organ in the flower with stigma, style, and ovary parts.

Pith The ground tissue composing the center of the primary plant body with vascular bundles arranged around it.

Plastid An organelle bound by a double unit membrane; e.g., chloroplasts, chromoplasts.

Pollen grain The male gametophyte of angiosperms and gymnosperms.

Polypeptide A chain of amino acids that comprise a protein.

Polyploidy More than two sets of chromosomes in a single cell.

Primary root First root emerging from a seed.

Primary tissues The tissues derived from the root and shoot apical meristems.

Procambium The tissue that eventually becomes the primary vascular tissue.

Prokaryote Organisms lacking a membrane bound nucleus or organelles.

Prophase The earliest stage of mitosis and meiosis, where chromosomes condense and the nuclear membrane disintegrates.

Proplastid Immature form of a plastid.

Protodermis The outer tissue that eventually becomes the epidermis.

Protoplast Refers to all the components of a plant cell except the cell wall.

Radicle Root of the embryo or developing seed.

Ribonucleic acid (RNA) A nucleic acid formed on the chromosome DNA that functions in protein synthesis.

Ribosomes RNA-containing bodies in the cytoplasm, mitochondria, and chloroplasts where protein synthesis occurs.

Saprophytes Plants that secure food by digesting nonliving organic matter.

Sapwood Outer wood that is light in color; active in water transport.

Scientific method A planned procedure for the collection of data based on controlled experiments and hypotheses.

Sclereid A sclerenchyma cell with thick secondary walls and with pits.

Sclerenchyma Specialized tissue for support; classified as sclereids or fibers.

Secondary cell wall The cell layer that is innermost within the cell.

Secondary roots Roots that develop from the pericycle of the primary root.

Seed coat The protective outer covering of the seed.

Senescence An irreversible period of plant decline; aging.

Shoot The aerial portion of a plant; stems and leaves.

Sieve area An area in the wall of sieve elements where pores can connect adjacent cell protoplasts.

Sieve cell Functions in food conduction.

Species A reproductively isolated interbreeding population.

Spongy parenchyma The bottom part of leaf mesophyll.

Stamen The male organ of the flower; parts are anthers and stamen.

Statoliths Starch grains in root cap cells.

Stoma (stomata) The opening and adjacent guard cells found in the epidermis of stems and leaves.

Style Structure that connects the stigma and ovary.

Suberin A fat substance deposited in the walls of cork cells.

Symbiosis The close relationship of two or more organisms of different species.

Syngamy Fusion of gametes to form a zygote.

Taproot system A root system with a significant, large main root.

Telophase The last stage of mitosis or meiosis; nuclei begin to reform and cytokinesis starts.

Thorn Modified branch; hard and pointed structure.

Tissue A group of cells that specialize to provide a function.

Tracheid An elongated cell that conducts water and serves in support.

Transpiration The evaporation of water away from the surface of the plant.

Trichomes Single or multicellular outgrowths of the epidermis.

Turgidity A condition where a cell is firm due to sufficient uptake of water.

Ultraviolet radiation Light outside the visible spectrum on the violet side.

Unicellular Composed of a single cell.

Vascular bundle A conducting element of phloem, xylem, and support tissue.

Vascular cambium Meristematic cells that divide to yield secondary xylem and phloem; diameter of plant will also increase.

Vein A vascular bundle found in a leaf.

Vessel Vessel cells or members connected end to end for food and water conduction.

Vessel member Xylem cells that are connected end to end to form a vessel.

Xylem Specialized tissue for the conduction of water and nutrients.

Zygote The cell that forms by the process of syngamy.

Appendix C: Directory of Contacts

Microscopic Analyses

For Training and Consultation

McCrone Research Institute
2820 S. Michigan Avenue
Chicago, IL 60616
Tel.: 312-842-7100
Fax: 312-842-1078
Website: www.mcri.org

Plant DNA Typing Informational Resources*

Dr. Heather Miller Coyle
Department of Public Safety
Division of Scientific Services
278 Colony Street
Meriden, CT 06451
Tel.: 203-639-6400
Fax: 203-639-6485
Email: C4ensic@yahoo.com

* In addition to the resources listed here, chapter and case contributors may be useful resources for forensic botany questions and casework considerations. It is also a good idea to check with local resources at state and private forensic laboratories as well as available academic resources from local universities.

0-8493-1529-8/05/$0.00+$1.50

Dr. Gary Shutler
Washington State Patrol
Crime Lab Division
2203 Airport Way South, Suite 250
Seattle, WA 98134-2027
Tel.: 206-262-6020
Email: gshutle@wsp.wa.gov

Dr. Robert A. Bever
The BODE Technology Group, Inc.
7364 Steel Mill Drive
Springfield, VA 22150
Tel.: 703-644-1200
Fax: 703-644-7730
Email: Robert.Bever@bodetech.com

Dr. Jon Wetton
The Forensic Science Service, Trident Court
Solihull Parkway
Birmingham Business Park
Solihull B37 7YN
Tel.: 44 (0) 121 329 5429
Fax: 44 (0) 121 622 2051
Website: www.fss.org.uk

Dr. Paul Keim
Northern Arizona University
Department of Biological Sciences
Flagstaff, AZ 86011
Tel.: 928-523-7515
Fax: 928-523-0639

Suppliers of Forensic Products and Equipment

Applied Biosystems
850 Lincoln Centre Drive
Foster City, CA 94404
Tel.: 1-800-345-5224
Website: www.appliedbiosystems.com

Promega Corporation
2800 Woods Hollow Road
Madison, WI 53711-5399
Tel.: 608-274-4330
Fax: 608-273-6989
Website: www.promega.com

MiraiBio Inc.
1201 Harbor Bay Parkway
Suite 150
Alameda, CA 94502
Tel.: 1-800-624-6176
Fax: 510-337-2099
Website: www.miraibio.com

MOBIO Laboratories Inc.
P.O. Box 606
Solana Beach, CA 92075
Tel.: 1-800-606-6246
Fax: 760-929-0109
Website: www.mobio.com

D^2 Biotechnologies Inc.
P.O. Box 78843
Atlanta, GA 30314
Tel.: 404-577-0919
Fax: 404-577-0497
Website: www.d2biotech.com

QIAGEN Inc.
28159 Avenue Stanford
Valencia, CA 91355
Tel.: 1-800-426-8157
Fax: 1-800-718-2056

Appendix D: Biographies

Dr. Henry C. Lee is one of the world's foremost forensic scientists. Dr. Lee's work is considered a landmark in modern-day criminal investigations. He has been a prominent player in many of the most challenging cases of the last 45 years. Dr. Lee has worked with law enforcement agencies in helping to solve more than 6000 cases. In recent years, his travels have taken him to England, Bosnia, China, Brunei, Bermuda, the Middle East, South America, and other locations around the world.

Dr. Lee's testimony figured prominently in the O.J. Simpson trial, and in convictions of the "Woodchipper" murderer as well as hundreds of other murder cases. Dr. Lee has assisted local and state police in their investigations of other famous crimes, such as the murder of JonBenet Ramsey in Boulder, Colorado; the 1993 suicide of White House Counsel Vincent Foster; the murder of Chandra Levy; the kidnapping of Elizabeth Smart; and the reinvestigation of the Kennedy assassination.

Dr. Lee is currently the Chief Emeritus for the Division of Scientific Services and was the Commissioner of Public Safety for the state of Connecticut from 1998 to 2000. He also served as Chief Criminalist for the state of Connecticut from 1979 to 2000. Dr. Lee was the driving force in establishing a modern State Police Forensic Science Laboratory in Connecticut.

In 1975, Dr. Lee joined the University of New Haven, where he created the school's Forensic Sciences program. He has also taught as a professor at more than a dozen universities, law schools, and medical schools. Though challenged with the demands on his time, Dr. Lee still lectures throughout the country and world to police, universities, and civic organizations. Dr. Lee has authored hundreds of articles in professional journals and has coauthored more than 30 books, covering the areas of DNA, fingerprints, trace evidence, crime scene investigation, and crime scene reconstruction. His recent books

0-8493-1529-8/05/$0.00+$1.50

Famous Crimes Revisited, *Cracking Cases*, and *Blood Evidence*, have been well received by the public.

Dr. Lee has been the recipient of numerous medals and awards, including the 1996 Medal of Justice from the Justice Foundation and the 1998 Lifetime Achievement Award from the Science and Engineer Association. He has also been the recipient of the Distinguished Criminalist Award from the American Academy of Forensic Sciences and the J. Donero Award from the International Association of Identification. In 1992 he was elected a distinguished Fellow of the American Academy of Forensic Sciences.

Dr. Lee was born in China and grew up in Taiwan. He first worked for the Taipei Police Department, attaining the rank of captain. With his wife, Margaret, Dr. Lee came to the U.S. in 1965, and he earned his B.S. in Forensic Science from John Jay College in 1972. Dr. Lee continued his studies in biochemistry at NYU where he earned his master's degree in 1974 and Ph.D. in 1975. He has also received special training from the FBI Academy, ATF, RCMP, and other organizations. He is a recipient of seven honorary doctorate degrees from universities in recognition of his contributions to law and science. Dr. and Mrs. Lee have been married for 40 years and have two grown children, a daughter, Sherry, and a son, Stanley.

Dr. Gary Shutler is the DNA Technical Leader for the Washington State Patrol Crime Laboratory Division and is based in the Seattle Laboratory. He has over 26 years of experience in forensic science. Previous positions include service with the Royal Canadian Mounted Police (RCMP) Forensic Laboratory System in the Ottawa and Winnipeg Laboratories. In the mid to late 1980's, he initiated the RFLP DNA typing program for the RCMP. During the mid to late 1990's, his work included validation studies for PCR-based STR analysis and casework applications. He has a doctorate degree from the University of Ottawa involving the genetic and physical mapping of the myotonic (muscular) dystrophy gene. This work helped lead to the identification of the genetic basis for the disease. Dr. Shutler has authored and coauthored numerous scientific publications including papers in *Nature*, *Science*, *Cell*, the *Canadian Journal of Forensic Science*, and the *Journal of Forensic Sciences.*

Dr. Robert A. Bever is the Director of Research at the Bode Technology Laboratory in Springfield, Virginia. Dr. Bever's research focuses on developing and implementing methods for human DNA identification, DNA identification of botanical material from trace evidence, and DNA extraction methodologies from soil and water.

He has directed several laboratories serving the human identity market since 1989. These laboratories have been involved in the analysis of individual criminal cases, the processing of felon databanking samples, and the analysis of thousands of parentage cases. He has testified in over 150 court cases, including several Frye and Daubert hearings. He has published numerous articles relating to the use of DNA typing for forensic applications. Additionally, he has spoken to a variety of audiences on the application of DNA typing in criminal and parentage investigations.

Dr. Bever received his Ph.D. in microbiology from the University of Maryland, where he performed research on the identification and characterization of *Vibrio* species from the environment and from clinical specimens. He performed postdoctoral research at the University of Rochester and the Oregon Health Science University. His primary research interest at that time was the elucidation of the pathogenic mechanisms relating to bacterial lung infections.

Dr. Lynne Milne has a Bachelor of Science degree with Honors in geology from the University of Western Australia, and a Ph.D. in palynology from the University of Queensland. Dr. Milne worked as a consultant palynologist while her family was young, and then won a Research Fellowship in the Department of Geography at the University of Western Australia. She is currently an Adjunct Research Fellow with the School of Earth and Geographical Sciences and the Centre for Forensic Science at the University of Western Australia. Dr. Milne has 20 years of experience in tertiary palynology and comparative pollen morphology, and has been involved in forensic palynology for the past seven years. At present, she teaches forensic palynology as a postgraduate course at the Centre for Forensic Science, participates in police education, and conducts casework for the Western Australian Police Service. Her other research interests include aerobiology, palynology of contemporary wetlands, and the comparative morphology of fossil and modern pollen of the plant family Proteaceae.

Dr. Dallas Mildenhall is employed by the Institute of Geological and Nuclear Sciences in New Zealand. He has a Doctoral Degree from Victoria University of Wellington in forensic and antarctic palynology. He has been involved in forensic palynology since 1973 and has worked on more than 170 forensic palynology cases, mainly relating to drugs, murder, and assault. He regularly offers courses in forensic palynology and has been an invited keynote speaker at the American Association of Stratigraphic Palynologists Conference. He has also offered a workshop and is involved in the organization of the Symposia on Forensic Palynology at the 17th International Symposium on the Forensic Sciences and the 11th International Palynological Congress. He has received six citations for his assistance in five cases where forensic palynology was integral to the final outcome.

He has been invited to give a course on forensic palynology at the University of Vienna and is a member of the Centre for International Forensic

Assistance. Internationally, he has worked, advised, or lectured in many countries, including Australia, England, Germany, India, Malaysia, Poland, The Netherlands, Thailand, and the U.S.

Major Timothy M. Palmbach is the current Commanding Officer and Director for the Division of Scientific Services within the Connecticut Department of Public Safety. In that capacity, he oversees the activities of three separate professional laboratories: Forensic Science Laboratory, Controlled Substance and Toxicology Laboratory, and Computer Crime and Electronic Evidence Laboratory. Also, Major Palmbach is a part-time faculty member at the University of New Haven Forensic Science Department.

Mr. Palmbach earned a juris doctor from the University of Connecticut, a Master of Forensic Science degree from the University of New Haven, and a Bachelor of Science in chemistry from the University of New Haven. He is a member of the Connecticut Bar Association and the Connecticut United States District Court Bar Association. He has authored and coauthored numerous scientific and professional publications, and is a coauthor of *Henry Lee's Crime Scene Handbook.* Also, he has testified in state and federal courts in Connecticut as well as other states as an expert witness, and provided testimony relevant to crime scene processing, blood spatter and stain interpretation, shooting incident reconstruction, and digital enhancement methods.

Paul J. Penders is a specialized (forensic) photographer at the Connecticut Forensic Science Laboratory. Since 1989, Mr. Penders has photographed numerous crime scenes and a variety of physical evidence, including latent fingerprints and other forms of impression evidence. Mr. Penders teaches forensic photography at the Henry C. Lee Institute of Forensic Science and is a frequent guest lecturer at other colleges and universities. Mr. Penders also serves on the executive committee of the Federal Bureau of Investigation-sponsored Scientific Working Group for Imaging Technology (SWGIT) and the Forensic Imaging Certification Board of the International Association for Identification. In addition to forensic imaging, Mr. Penders has performed commercial photography for over 20 years with a client list that includes Yale University, the Hospital of St. Raphael, and Wesleyan University.

Dr. Carll Ladd obtained his Ph.D. in molecular biology and genetics from the University of Connecticut in 1990 and conducted postdoctoral DNA research at St. Francis Hospital, Hartford, Connecticut. Dr. Ladd is a Supervising Criminalist of the Forensic Biology–DNA section at the Department of Public Safety, Division of Scientific Services Forensic Laboratory in Meriden, Connecticut. He is an adjunct assistant professor at the University of Connecticut and is currently the representative for the Connecticut State Forensic Laboratory for SWGDAM

(Scientific Working Group for DNA Analysis Methods). Dr. Ladd has coauthored many book chapters and articles in the field of forensic DNA analysis. His research interests include DNA mixture interpretation and the development of new DNA typing technologies. He has worked on numerous high-profile cases, including Vincent Foster and *Connecticut v. Torres*, Aillon.

Elaine Pagliaro is currently the Assistant Director at the Connecticut Department of Public Safety, Division of Scientific Services Forensic Science Laboratory, where she has worked in criminalistics and forensic biology for more than 22 years. During that time, Pagliaro has been involved in most of the major criminal investigations in Connecticut and in cases of national prominence, such as the investigation of the death of Vincent Foster and the infamous "Woodchipper" murder case. She has qualified as an expert witness in the areas of forensic biochemistry, DNA, hair analysis, and crime scene reconstruction in Connecticut, Illinois, New York, Pennsylvania, Rhode Island, Vermont, and Louisiana. She is a fellow of the American Academy of Forensic Sciences, a member of several other professional organizations, and serves on the Connecticut Commission for the Standardization of the Collection of Sexual Assault Evidence. Pagliaro is an adjunct faculty member at several Connecticut universities and colleges, where she instructs students on criminalistics, nursing, and the law. A *summa cum laude* graduate of the Quinnipiac College School of Law, Pagliaro is a member of the bar in Connecticut, New Hampshire, and the Federal District of New Hampshire.

Nicholas Yang is currently employed as a criminalist in the DNA section at the Connecticut Department of Public Safety Forensic Science Laboratory. Previous employment includes a position at the Armed Forces DNA Identification Laboratory, specializing in mitochondrial DNA testing. In addition to the 8 years of experience in forensic DNA science, he is pursuing a Ph.D. degree in population genetics from the University of Connecticut in Storrs. Mr. Yang has a Master's of Forensic Science degree from George Washington University and a Bachelor of Science in biochemistry from the University of Maryland, College Park. He has interactions with many crime scene investigators and scientists from foreign countries, such as Taiwan and Croatia, in discussing forensic science issues. Mr. Yang has coauthored numerous scientific articles in journals such as the *Journal of Forensic Science and Croatian Medical Journal*. His research interests include DNA mixture analysis and applications of novel DNA techniques.

Eric Carita, born in Okinawa, Japan, and was raised in Thompson, Connecticut, as the youngest of three children. Eric received his Bachelors degree in sociology with a focus in criminology from Eastern Connecticut State University and his Master of Science in genetics from the University of Connecticut. His graduate research involved Y chromosomal STR subpopulation validation and AFLP typing of marijuana for forensic applications. Employed by the Henry C. Lee Institute of Forensic Science at the University of New Haven, Carita works at the Connecticut State Forensic Science Laboratory where he is creating a state, national, and international marijuana AFLP-DNA database. Eric lives in Thompson, Connecticut, with his wife, Wendi, and his two children, Christian and Emily.

Dr. Ian C. Hsu received his Ph.D. in physics from the University of Wisconsin–Madison in 1989 and has spent a number of years working at the Stanford University SLAC, Accelerator Department. Since 1998, he has been a professor and director of the Nuclear Science Department at National Tsing Hua University, where his work focused on the design and development of biochip systems. Those topics involved "The Development of HHHM DNA/Protein Microarray," "The Construction and Improvement of Microarrays for Studies of Gene Expression International Collaboration," and "The Study of Biological Effects and the Gene Expression Pattern of EM Radiation by Gene Chips."

Cheng-Lung Lee received his Bachelor and Master degrees of forensic science from Central Police University at R.O.C. (Taiwan) in 1993 and 1998, respectively. He has spent 10 years working in the Hsin-Chu Municipal Police Bureau, Criminal Investigation Department, and the Forensic Science Division, where his work focused on crime science investigation and the development of new techniques for forensic science. Since 2000, he entered National Tsing Hua University, Department of Nuclear Science, for further study in a new technique to be applied to forensic science. He was selected to be directorate of the Taiwan Forensic Science Association and passed the Ph.D. qualifying exam in 2002. He won a scholarship for a "Graduate Students Study Abroad Program" from the National Science Council, R.O.C. (Taiwan), and then went abroad to the U.S. Cheng-Lung Lee is a research scholar at the University of New Haven and is a visiting scientist at the Connecticut Department of Public Safety, Division of Scientific Services. For his doctoral research, he is developing a novel method for the DNA analysis of digested seeds in forensic samples.

James Chun-I Lee received B.A. and M.A. degrees in law from the Central Police University, Taiwan (1979 and 1984), an M.S. degree in forensic science from the University of New Haven, Connecticut (1988), and a Ph.D. degree in forensic science from the University of Strathclyde, Scotland (2001). Currently, Dr. Lee is a professor in the Department of Forensic Science at Central Police University.

Prior to entering academia, Dr. Lee worked as a police officer and criminalist in the police department in Taiwan for 10 years. His current duties include teaching and conducting research in forensic biological analyses. Since 1990, Dr. Lee has published more than 50 peer-reviewed articles in blood group genotyping, genetic diversity studies on STR and mitochondrial DNA markers, DNA analysis on endangered species, STR isolation of species related to forensics, and development of new methods in biological evidence analysis. Currently, he is the vice president of the Taiwan Academy of Forensic Science.

Dr. Hsing-Mei Hsieh received her Ph.D. in life sciences from National Tsing Hua University (Hsinchu, Taiwan) in 1996. Dr. Hsieh is currently an associate professor at the Forensic Science Department of Central Police University in Taiwan. Dr. Hsieh has directed some plans financially supported by the National Science Council of Taiwan in botanic identifications, especially in forensically relevant plants. The topics of these plans are "Application and Evaluation of Botanic DNA Identification to Criminal Investigation," "Species Identification by Chloroplast DNA and Nuclear RNA Genes in Automated Approaches of Forensic Botanic Identification," "Study of STR Cloning on Forensic Plants," and "The Study of Phylogenetic Relationships of Orchids by DNA Analysis in Forensic Applications — The Study of *Phalaenopsis amabilis* as the Model System." The results of these studies were published in international journals. In addition to the botanical identifications, Dr. Hsieh engages in the identification work of conservation animals, such as rhinoceroses.

Dr. Adrian Linacre read zoology at Edinburgh University (Scotland) before studying toward a Doctorate of philosophy at Sussex University (England) in molecular genetics. Since 1994, he has been employed at the Forensic Science Unit of Strathclyde University in Glasgow (Scotland), where he consults on criminal and civil cases for the courts in the U.K. and is involved with many international cases. He is an assessor for the Council for the Registration of Forensic Practitioners and is a Registered Forensic Practitioner in the area of human contact traces (DNA and body fluids). His research focuses on the use of DNA in nonhuman studies, particularly plant and wildlife studies.

Dr. Margaret Sanger received her Ph.D. in plant physiology from the University of California at Davis in 1987 and has spent a number of years working in agricultural biotechnology in both academic and industry settings. Her work has focused on the design and development of gene expression systems applied to plants, foods, and microorganisms. Since 1999, she has worked at the Kentucky State Police Central Forensic Laboratory in the Appalachian H.I.D.T.A. Signature Laboratory, where she is currently developing methods for the forensic identification and genetic profiling of marijuana. She is also involved in human DNA analysis methods.

Dr. Bruce Budowle received a Ph.D. in Genetics in 1979 from Virginia Polytechnic Institute and State University. From 1979 to 1982, Dr. Budowle was a postdoctoral fellow at the University of Alabama at Birmingham. Working under a National Cancer Institute fellowship, he carried out research predominately on genetic risk factors for such diseases as insulin dependent diabetes mellitus, melanoma, and acute lymphocytic leukemia.

In 1983, Dr. Budowle joined the research unit at the FBI to carry out research, development, and validation of methods for forensic biological analyses. The positions he has held at the FBI include: research chemist, program manager for DNA research, and Chief of the Forensic Science Research Unit, and currently, he is the only Senior Scientist for the entire Laboratory Division of the FBI. Dr. Budowle has contributed to the fundamental sciences as they apply to forensics in analytical development, population genetics, statistical interpretation of evidence, and in quality assurance. Some of the methods he developed are: 1) analytical assays for typing a myriad of protein genetic marker systems, 2) designing electrophoretic instrumentation, 3) developing molecular biology analytical systems to include RFLP typing of VNTR loci and PCR-based SNP assays, VNTR and STR assays, and direct sequencing methods for mitochondrial DNA, and 4) designing image analysis algorithms. Dr. Budowle has worked on laying some of the foundations for the current statistical analyses in forensic biology and defining the parameters of relevant population groups. He has published more than 350 articles, made more than 390 presentations (many as an invited plenary speaker), and testified in well over 200 criminal cases in the areas of molecular biology, population genetics, statistics, quality assurance, and forensic biology. In addition, he has authored or coauthored books on molecular biology techniques, electrophoresis, and protein detection. Dr. Budowle has been directly involved in developing quality assurance (QA) standards for the forensic DNA field. He has been a member of the Scientific Working Group on DNA Methods for 14 years (of which he was chair for six), Chair of the DNA Commission of the International Society of Forensic Genetics, and a member of the DNA Advisory Board. He was one of the architects of the CODIS National DNA database, which maintains DNA profiles from convicted felons, from evidence in unsolved cases, and from missing persons.

Some of Dr. Budowle's more recent efforts are in counter terrorism, primarily in identification of victims from mass disasters and in efforts involving microbial forensics. Dr. Budowle is an advisor to New York State in the effort to identify the victims from the WTC attack. He assisted Celera in establishing a mitochondrial DNA sequencing program to enable high throughput sequencing of remains from the WTC. Bioterrorism is a major concern for U.S. security. Dr. Budowle is one of the lead individuals in the forensic applications on bioterrorism and is developing a new field known as microbial forensics. In the area of microbial forensics, Dr. Budowle is the current chair of the Scientific Working Group on Microbial Genetics and Forensics, whose mission is to set QA guidelines, develop criteria for biologic and user databases, set criteria for a National Repository, and develop forensic genomic applications for the National Biodefense Analysis and Countermeasures Center at Ft. Dietrick. He also has served on the Steering Committee for the Colloquium on Microbial Forensics sponsored by American Society of Microbiology and was the organizer of two Microbial Forensics Meetings at The Banbury Center in Cold Spring Harbor Laboratory.

Index

W

X

Y

Z